Current Topics in Neuroendocrinology Volume 6

This collection of studies was conceived as part of a two volume review of the subject. The contents of Volume 4 are listed below.

Neurobiology of Vasopressin

Neurobiology of Oxytocin

Editors
D. Ganten and D. Pfaff

Contributors
J. P. H. Burbach M. L. Forsling R. Ivell
G. L. Kovács B. T. Pickering D. G. Porter
I. C. A. F. Robinson R. W. Swann D. C. Wathes

With 38 Figures

Springer-Verlag
Berlin Heidelberg New York Tokyo

Editors

Dr. DETLEV GANTEN, M.D., Ph.D.
Pharmakologisches Institut
Universität Heidelberg
Im Neuenheimer Feld 366
6900 Heidelberg/FRG

Dr. DONALD PFAFF, Ph.D.
Rockefeller University
York Avenue, and 66th Street
New York, NY 10021/USA

The picture on the cover has been taken from Nieuwenhuys R., Voogd J., van Huijzen Chr.: The Human Central Nervous System. 2nd Edition. Springer-Verlag Berlin Heidelberg New York 1981

ISBN-13: 978-3-642-70416-1 e-ISBN-13: 978-3-642-70414-7
DOI: 10.1007/978-3-642-70414-7

Library of Congress Cataloging in Publication Data. Neurobiology of oxytocin. (Current topics in neuroendocrinology; v. 6). Includes bibliographies and index. 1. Oxytocin. 2. Neurophysiology. I. Ganten, D. (Detlev), 1941– . II. Pfaff, Donald W., 1939– . III. Burbach, Johannes Peter Henri, 1954– . IV. Series. [DNLM: 1. Neurobiology. 2. Oxytocin. W1 CU82Q v. 6/QV 173 N494] QP572.O9N48 1986 612′.63 86-4006

Softcover reprint of the hardcover 1st edition 1986

2121/3130-543210

Contents

Biosynthesis of Oxytocin in the Brain and Peripheral Organs

R. IVELL

Contents

1 Introduction

Historically, oxytocin biosynthesis has always been associated with the synthesis of the chemically related hormone vasopressin. Although the two hormones pursue quite different functions in the mammal, oxytocin being primarily associated with smooth muscle contraction of the female reproductive system and vasopressin with vasopressor and antidiuretic activities, two types of evidence have linked these hormones together.

The first type of evidence is anatomical. Both hormones were shown to be released from the posterior pituitary, where they existed in association with what later became identified as neurophysins, cysteine-rich proteins of about 10000 molecular weight. The relationship between these molecules was very confusing as long as pituitary acetone powders, in the absence of modern peptide separation techniques, were used as a source of the hormones. At one time it was considered that the neurophysins, oxytocin, and vasopressin formed parts of a larger protein – the Van Dyke protein – with a molecular weight of about 30000 (Van Dyke et al. 1942). With hindsight we can see that this assumption arose on the one hand,

Institut für Zellbiochemie und Klinische Neurobiologie, Universität Hamburg, D-2000 Hamburg

Current Topics in Neuroendocrinology, Vol. 6

because of specific interactions between the nonapeptide hormones and the neurophysins on a 1 : 1 stoichiometric basis; and on the other hand, because of interactions between neurophysin molecules, which readily and specifically dimerize, especially when linked with a nonapeptide hormone (Breslow 1979; Angal and Chaiken 1982). Clarification of this point was further aggravated by the incidental tendency in the two most commonly used experimental animals, the cow and the rat, for oxytocin and vasopressin to be produced in approximately equimolar quantities (Acher 1979). This is not the case in the human or the whale (Acher 1979; George 1978), where vasopressin is present in a greater amount.

The second type of evidence was evolutionary. Chemical analysis of vasopressin- and oxytocin-like peptides throughout the vertebrates pointed to there being two essential peptide types, reflecting the mammalian molecules, which probably arose from a common ancestral hormone in the cyclostomes about 400 million years ago (Acher 1979).

Now that the biosynthetic pathways for both hormones have been elucidated using a combination of in vivo and in vitro techniques, not the least important of which is recombinant DNA methodology, it can be seen that most of the early difficulties arose because oxytocin and vasopressin are synthesized from very similar but independent genes, via a comparable series of intermediate steps, along closely parallel neuroanatomical routes.

2 Sites of Synthesis

All the early studies of oxytocin biosynthesis focused on the classic hypothalamo-neurohypophyseal system. The Gomori's stain, which appears to be specific for the cysteine-rich contents of secretory granules, showed that the nerve terminals of the posterior pituitary were probably being fed via long axons from cells of the anterior hypothalamus (Bargmann and Scharrer 1951). Subsequently, immunohistochemistry identified two magnocellular hypothalamic nuclei – the supraoptic nucleus (SON) and the paraventricular nucleus (PVN) – as well as some parvocellular neurones in the caudal PVN which specifically contain oxytocin and its related neurophysin I in their cell bodies (Dierickx and Vandesande 1979; Sofroniev 1983, 1985). To date, unlike vasopressin neurones, oxytocin-producing cell bodies appear to be absent in nonhypothalamic brain areas (Sofroniev 1985). The hypothalamic neurones do, however, project into several extra-hypothalamic regions, as well as into the median eminence and posterior pituitary (Sofroniev 1983, 1985).

Outside the brain, oxytocin-containing cells have been identified in several different tissues, including the adrenal medulla (Ang and Jenkins 1984; Nicholson et al. 1984), the testicular interstitial cells (Nicholson et al. 1984; Guldenaar and Pickering 1985), the corpus luteum (Wathes and Swann 1982; Flint and Sheldrick 1982; Fields et al. 1983), and the placenta (Fields et al. 1983; Makino et al. 1983), and in small cell carcinomas of the lung (Maurer et al. 1983). These identifications are based on immunohistochemistry, radioimmunoassay, HPLC, or a combination of these; but, except for the corpus luteum, chemical characterization by sequence analysis either of the peptide or its mRNA is still lacking (Ivell and Richter 1984 a).

Without exception, all studies in recent years have emphasized that oxytocin and its neurophysin I are produced in a population of cells discrete from those producing vasopressin and its associated neurophysin II (Dierickx and Vandesande 1979; Sofroniev 1983). No cells have yet been identified in which vasopressin and oxytocin are colocalized, although either of the hormones can share cells with other peptides, e.g., met-enkephalin, leu-enkephalin, or corticotropin-releasing factor (Martin and Voigt 1981; Whitnall et al. 1983; Wolfson et al. 1985).

3 Demonstration of Oxytocin Synthesis In Vivo

Earlier in vivo experiments were carried out on the vasopressin synthesizing system (Sachs et al. 1969). Always implicit in these studies was the notion that whatever was described for vasopressin was probably valid for oxytocin as well. This work led to the idea of there being a common precursor for the nonapeptide hormone and the neurophysin with which it is found associated in the neurosecretory granules of the posterior pituitary. Direct evidence that this was indeed the case for oxytocin first came in the seventies. It was shown that injection of (^{35}S)cysteine near the supraoptic nucleus of the rat hypothalamus is incorporated first into longer precursor proteins with a molecular weight of about 20000. It was further determined that these proteins, as they were transported axonally through the pituitary stalk to the neurohypophysis, give rise via proteolytic cleavage to molecules which could be identified as vasopressin, oxytocin, and neurophysins (Gainer et al. 1977). Only at a later date was it possible for the oxytocin nonapeptide to be identified within the large hypothalamic precursor molecule, and to be set free from this precursor by treatment with trypsin (Russell et al. 1980). More recently, HPLC techniques have been refined such that all the precursors and products of vasopressin and oxytocin biosynthesis present in the hypothalamus can be identified in a single chromatographic run (Swann et al. 1982), thus simplifying the analysis of oxytocin in vivo biosynthesis and its regulation.

The biosynthesis of extra-hypothalamic oxytocin has been demonstrated in vivo only in the case of the bovine corpus luteum (Swann et al. 1984). The nonapeptide appeared to be the cleavage product of a larger hormone precursor, which also included the neurophysin moiety, thus conforming to the pattern established from the hypothalamus.

The in vivo synthesized oxytocin precursor does not bind to the mannose-specific lectin concanavalin A, indicating that it is not glycosylated (Russell et al. 1980). Although definite information on other protein modifications is lacking, it does not appear to be phosphorylatable (McKelvy 1975), and, except for the hormone amidation (see below), the end products are equally unmodified.

4 Cell-Free Translation of mRNA

In vivo data showed that oxytocin was synthesized (a) in a ribosome-dependent manner, encoded therefore by a messenger RNA transcript of a putative oxytocin gene; (b) with at least one other peptide moiety, its associated neurophysin, as part of a common precursor. More detail on the structural organization of the nonapeptide common precursors was obtained by cell-free translation of bovine hypothalamic mRNA in wheat germ and reticulocyte lysate derived in vitro systems (Schmale et al. 1979; Schmale and Richter 1980). Cell-free translation yields the primary preprohormones which precede all the subsequently modified forms found in vivo. Using specific antisera raised against either oxytocin, vasopressin, neurophysin I, or neurophysin II, two common precursors were identified in these in vitro systems: one the precursor to vasopressin-neurophysin II, the other to oxytocin-neurophysin I.

Cotranslational addition of microsomal membranes cleaves the signal peptide at the N-terminus of the preprohormone, and causes n-glycosylation of appropriate -Asn-X-Thr/Ser- sites within the protein. Bovine preprooxytocin (molecular weight 16 500) was thus shown to have a short N-terminal signal sequence, but unlike preprovasopressin no site for glycosylation, the membrane-supplemented system yielding a pro-oxytocin with molecular weight of 15 500 (Schmale and Richter 1980).

Allochthonous mRNA can also be translated in the microinjected frog oocyte. Not only was it possible to identify a similar pro-oxytocin of about 15 000 molecular weight within the oocyte, but this protein was also specifically secreted, apparently without further cleavage, into the incubation medium (Richter 1983). This experiment demonstrated that the pro-oxytocin molecule includes all the information necessary to secure its specific packaging into granules, transport, and secretion. Apparently, the frog oocyte does not possess the specific proteases needed to cleave the nonapeptide from its accompanying neurophysin (Richter 1983).

5 Oxytocin Precursor Structure Elucidated by Recombinant DNA Methods

Since it was shown that hypothalamic mRNA was sufficiently intact to allow the in vitro translation of the nonapeptide hormone precursors, the same RNA was also used to prime cDNA synthesis for a recombinant DNA library. In the course of screening such a hypothalamic cDNA library for vasopressin-encoding clones, a second series of clones were identified which were partly homologous to the vasopressin cDNA. Sequencing of these clones yielded the primary nucleotide sequence for the oxytocin mRNA (Land et al. 1983). Its organization fully confirmed the supposition obtained by in vivo and in vitro experimentation: oxytocin nonapeptide was synthesized as a common precursor together with its specific neurophysin. Except for a short signal peptide at the N-terminus no other peptide moiety was encoded; thus unlike the vasopressin precursor (Land et al. 1982) there was no C-terminal glycopeptide. The bovine cDNAs encoding the vasopressin and oxytocin precursors were then used to isolate the respective genes from both bovine (Ruppert et al. 1984) and rat (Ivell and Richter 1984b) genomic libraries.

5.1 Oxytocin Gene Structure

In both species so far studied, the protein-coding region of the oxytocin gene is divided into three exons by two short intervening sequences (Fig. 1). In the 5′ untranslated region upstream of the first exon, the so-called promoter region, there is a typical RNA polymerase II binding site, CATAAA, 68 nucleotides before the first methionine codon. S1 nuclease mapping of this region in the rat gene (Ivell and Richter 1984b) showed that transcription starts at either of the three adenosine residues 26–28 bases following this sequence. A comparison between the bovine and rat gene sequences (Ivell and Richter 1984b; Ruppert et al. 1984) further in the 5′ direction indicates several blocks of homology suggesting that specific regulation may occur via proteins binding at these points. In the vasopressin gene the corresponding region is very different, though highly homologous between species. This is in agreement with the independent regulation of the vasopressin and oxytocin genes, probably through the binding of hormone-specific promoter molecules in the 5′ regions.

The first exon encodes the predicted short signal peptide of 19 amino acids, an alanine evidently marking the point of signal peptidase cleavage. Oxytocin follows immediately after, such that it must lie at the extreme N-terminus of the common precursor found in vivo. The nonapeptide is separated by the sequence -Gly-Lys-Arg- from the first nine amino acids of neurophysin I encoded by the remainder of the first exon. The glycine of this triplet is known to provide the nitrogen for the amidation of the immediately preceding hormone C-terminus (Bradbury et al. 1982). The pair of basic amino acids, lysine-arginine, conforms to the typical cleavage signals now identified in the specific proteolysis of many polyprotein hormone precursors (Richter 1983).

Exon B encodes the central 67 amino acids of neurophysin I, exon C the remaining 17 residues as well as a supernumerary basic amino acid at the C-terminus immediately before the stop codon. In the rat this penultimate codon is for arginine, in the calf for histidine. No basic amino acid has been identified in this position by protein sequencing of the extracted neurophysins, and it may mark a further specific post-translational cleavage, such as has also been noted for other polyprotein precursors, e.g., corticotropin releasing factor (Furutani et al. 1983).

In the 3′ untranslated region, the bovine gene presents a single typical polyadenylation signal of the type AATAAA. In the rat, however, there are three repeated signals of this type, preceded by the less usual polyadenylation signal ATTAAA. Sequence analysis of rat hypothalamic cDNA clones indicates addition of the poly(A) tail at base 837 (Fig. 1, arrow) downstream of all polyadenylation signals (Ivell and Morley unpublished).

```
-354                                                  GATCCACGTAAGGGCTTCCTCACCCACTTGCGCTATCCAGCCTCATTCTGAGGT
-300 ATTGGATTTCTATGAAAAAACAGCTCTTGGCTAGGCTGCACCTCTACCCCCTCCAAGTCTCTTTATCCTCTTGTAGCTTAGGCCTCCCCTTCTAGGCTGT
-200 GTCCCTTTTGAGCTCAGGTCATTAGCTGAGGCGGTGACCTTGACCCCAGCCCAGACCCTGCAAATGAAGGGCCTGCTTCTAAACAGTGTGGAACAGTTTG
-100 ACCCAAGAGACCTGGCTGTGACCAGTCATGCAGTCACCCTCTTAGACTGGGCCCCACCATGGCAGTGGACAAGGCATAAAAAGGTCGGTCTGGCTGGAG

                                             MetAlaCysProSerLeuAlaCysCysLeuLeuGlyLeuLeuAlaLeuThrSerAlaC
   1 AAACCATCACCGACGGTGGATCTCGGACTGAACACCAACGCCATGGCCTGCCCCAGTCTCGCTTGCTGCCTGCTTGGCCTACTGGCTCTGACCTCCGCCT

     OXYTOCIN
     ysTyrIleGlnAsnCysProLeuGlyGlyLysArgAlaAlaLeuAspLeuAspMetArgLys
 101 GCTACATCCAGAACTGCCCCCTGGGCGGCAAGAGGGCTGCGCTAGACCTGGATATGCGCAAGGTGAGTCTCCCCGACGCCATCCCGTCCCGTTCTGGCAA

 201 GGCTAAGGACCAGAGATGCTCTCCCACCTTCAGAGAGCATCCCCTCACACTTGCCAGCCCTACGCGACCTCGCATGGGAGACCCGGAGCTCCCTCTGACG

                                                                                       CysLeuProCysGlyPro
 301 CCGTGAAGGCACGCTGTCCCTGCCCCACCACAGTCCCGATATGGATGGGGGCAGCGCCCATGCGCGTGTTTCCCCCCGCAGTGTCTTCCTTGCGGACCC

                                           NEUROPHYSIN
     GlyGlyLysGlyArgCysPheGlyProSerIleCysCysAlaAspGluLeuGlyCysPheValGlyThrAlaGluAlaLeuArgCysGlnGluGluAsnT
 401 GGCGGCAAAGGGCGCTGCTTCGGGCCGAGCATCTGCTGCGCGGACGAGCTGGGCTGCTTCGTGGGCACCGCCGAGGCGCTGCGCTGCCAGGAGGAGAACT

     yrLeuProSerProCysGlnSerGlyGlnLysProCysGlySerGlyGlyArgCysAlaThrAlaGlyIleCysCysSerProA
 501 ACCTGCCCTCGCCCTGCCAGTCTGGCCAGAAGCCTTGCGGAAGCGGAGGCCGCTGCGCCACCGCGGGCATCTGCTGTAGCCCGGGTGAGCAGGAGGGGCC

                                                                        spGlyCysArgThrAspProAlaCysAspP
 601 TAGCAGCAGGACCGACCGGCAGGGAGCCGTCGGGTTTGCTGCTCAGGCCACTCACCCATTTCTCTTGCAGATGGCTGCCGCACCGACCCCGCCTGCGACC

     roGluSerAlaPheSerGluArgEnd
 701 CTGAGTCTGCCTTCTCCGAGCGCTGAGCCCGCTTTGTGATGATACCTTTAGGGCGCTTCCTTCATTCCCCATGGCCACTACCAGAAAAAAATTAAAAAAA
 801 AAACATAAATAAATAAATAAAGCAGATTTCCTTTTCAAACTTGACTGGCGTCTAATTGTCAGAAACGGGAGGGAGGAGGAAAGGCACCGGGAACGCCATG
 901 GACATTGGCAATTCAGAGGGAGAACAGCCCAGCAGGCTGCTAGGAGGACCGAGGGGTCTCTACAGGTTGTCATCTCCAGTTAGGCCCGTTTCCCACTCCC
1001 GACTGGGGTAAGCCTGCGGCAGACAGACTTAGGAGCAGCATTTGGAAGTCCAAACAGCAATGGTGAACTTCTGTGGGAAAGGGCAAAGCCAAACAGGGGC
1101 CTGGTGATGGGAAGGCAGGCGAGGCGGCAAGGGACTCATCCAGTCACTTCCATATCTGGAGGAGACCTTGGCGGCGACCTCAGAGGTTGAGGGGGTAATC
1201 CCGGCCTAGAAGAAAGGGTCTTGGCCAGGGAGGGTAGTTGCAGAACCATAAGGTTTCTACAGAGCTGGGTT
```

Fig. 1. DNA sequence of the complete oxytocin gene from the rat. The *boxed* regions indicate 5′ sequence homologies between rat and cow which are possibly involved in gene regulation. *Arrows* indicate transcription start (nucleotides 1–3) and polyadenylation (nucleotide 837) sites

5.2 Comparison of the Vasopressin and Oxytocin Genes

Although it is clearly evident from a comparison of both amino acid and nucleotide sequences that the oxytocin and vasopressin genes derive from a common ancestor by gene duplication, the distribution of the homology within the genes requires additional comment (Fig. 2). Exon A exhibits approximately 70% homology, a figure concomitant with the age of the gene duplication and the conservation pressures to which the genes are subjected. Exon B, in stark contrast, is remarkably homologous – in the cow there is not a single nucleotide difference for 194 bases of the exon and continuing for a further 135 bases into the first intron. In the rat it is less marked though still noteworthy, with 134 bases containing only a single nucleotide change in the middle. Such homology is not consistent with an ancient gene duplication alone. Nor can it be attributed to strong conservation by selection at the protein phenotype level, since then some silent changes in the third position of a number of codons would be expected. In similar, though less pronounced examples this phenomenon has been attributed to a process of gene conversion (Baltimore 1981), whereby at the level of the DNA (since introns are also included) the two nonallelic genes for vasopressin and oxytocin are made more similar. Postulated is either a sort of mutual correction involving a new type of DNA repair mechanism, or an unequal recombination event. The relevance of this mechanism for the organism is not clear yet.

Exon C exhibits the least homology between the two genes. Significantly, only the codon for the supernumerary basic amino acid (oxytocin gene) is entirely homologous with the equivalent codon for the cleavage signal separating neurophysin and the C-terminal glycopeptide (vasopressin gene). Evolution has favored a strong divergence within this exon. This feature can be usefully exploited since DNA probes derived from this region of the gene or cDNA are exclusively specific for the oxytocin or vasopressin genes, whereas probes derived from other regions show greater or lesser extents of cross-reactivity. 3′ specific probes are therefore invariably used to measure levels of mRNA in various tissues, reflecting the activity there of the vasopressin or oxytocin genes.

Analysis of genomic DNA from male rats and calves which has been cleaved by various restriction enzymes, electrophoresed, transferred to nitrocellulose membranes, and probed with the 3′ probes specific for the oxytocin gene indicates only a single copy of this gene in each species (Ivell and Richter 1985). Therefore, detection of oxytocin synthesis in different tissues under different physiological conditions implies that this one gene is being transcribed in a tissue-specific and physiologically relevant manner.

For the vasopressin gene a natural experiment in the cow has shown that both alleles are expressed equally. Bovine neurophysin II exhibits a polymorphism at amino acid 89; in a mixed pituitary extract from many cows, 70% of neurophysin II has isoleucine in this position, the remaining 30% valine. But in individual pituitaries the proportion of each morph found is always 100%, 50%, or 0%, corresponding to the classic Mendelian distribution (Moore and Hope 1985). Whether both alleles of the oxytocin gene are similarly expressed has yet to be determined.

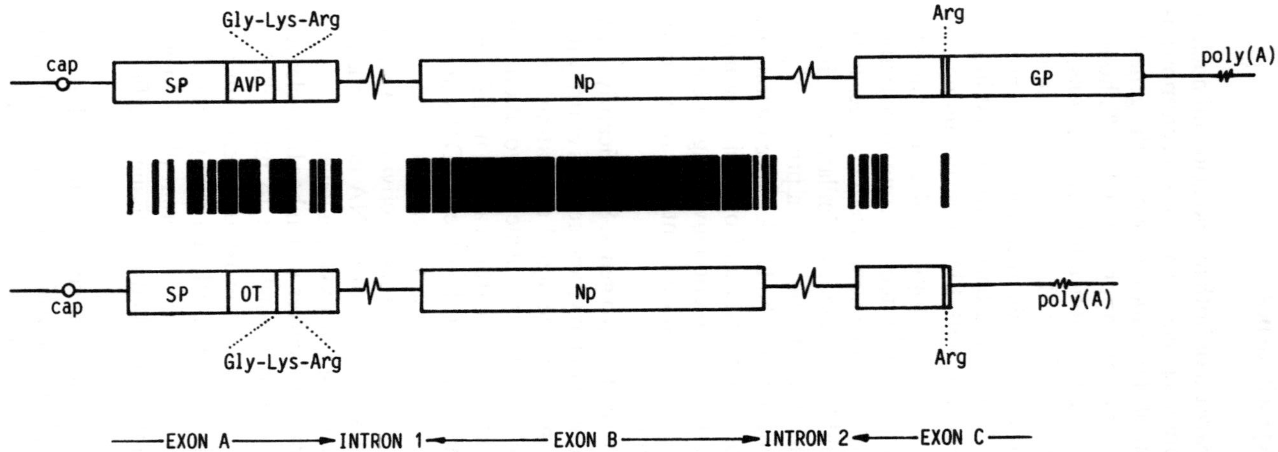

Fig. 2. Comparison of the vasopressin (*upper*) and oxytocin (*lower*) gene organization. *SP*, signal peptide; *AVP*, arginine vasopressin; *OT*, oxytocin; *NP*, neurophysin; *GP*, glycopeptide; *cap*, putative transcription start site; *poly A*, signal for mRNA polyadenylation; *black bars*, regions of nucleotide homology comprising 3 or more successive bases. Introns which are compressed in this figure show no apparent homology

6 Post-Translational Processing

Now that gene analysis has provided the complete primary amino acid sequence of the common precursor to oxytocin and neurophysin I, it is possible to interpret more precisely the events recorded in the in vivo and in vitro synthesis studies (Fig. 3). The preprooxytocin molecule predicted by the mRNA is only found in in vitro translation systems. Already while translation is proceeding the hydrophobic signal sequence recognizes the membrane of the endoplasmic reticulum, traverses this, pulling the nascent polypeptide chain from the now membrane-attached ribosome into the reticular lumen. Here, signal peptidase cleaves off the leader peptide, presumably at Ala_{19}. The remaining pro-oxytocin, which is not glycosylated, passes to the Golgi apparatus, where it is packaged into neurosecretory vesicles. En route between the Golgi and the nerve terminals of the neurohypophysis, specific enzymes within the granules further cleave the precursor to yield neurophysin I and oxytocin, the latter being simultaneously amidated. This post-translational processing is probably time rather than place dependent, since blockage of axonal transport by colchicine increases the proportion of end product to precursor in the hypothalamic cell bodies (Camier et al. 1985). The converse appears to be the case in the microinjected frog oocyte. Here evidently the processing peptidase(s) is (are) absent, but nonetheless packaging, movement of granules to the cell surface, and specific export proceed normally (Richter 1983).

It should be noted that although most neurophysin antisera will also recognize the precursor pro-oxytocin, this is rarely so for antioxytocin antibodies. These generally require the amidation for recognition and thus cannot recognize the nonapeptide before it is cleaved from the precursor. Thus colchicine treatment may improve immunohistochemical reactivity in cell bodies, not simply by blocking transport, but also by encouraging premature cleavage of the precursor in the perikarya.

7 Regulation of Oxytocin Gene Transcription in the Hypothalamus

The two classic functions of hypophyseal oxytocin, uterine contraction and mediation of the milk let-down reflex make it necessary to ask whether such responses are related only to increased release of the stored hormone from the pituitary, or whether there is a corresponding change in the rate of oxytocin neosynthesis? Although biosynthesis of oxytocin per se has not yet been looked at, the availability of specific DNA probes can be used to estimate the levels of oxytocin mRNA in the hypothalamus. This serves as a useful index of the transcription rate.

Using the technique of in situ hybridization illustrated in Fig. 4, McCabe et al. (1985) examined the cells of the supraoptic nucleus whose oxytocinergic neurones all project into the neurohypophysis. Silver grain counts of autoradiographed thin sections which had been hybridized to radioactively labelled, oxy-

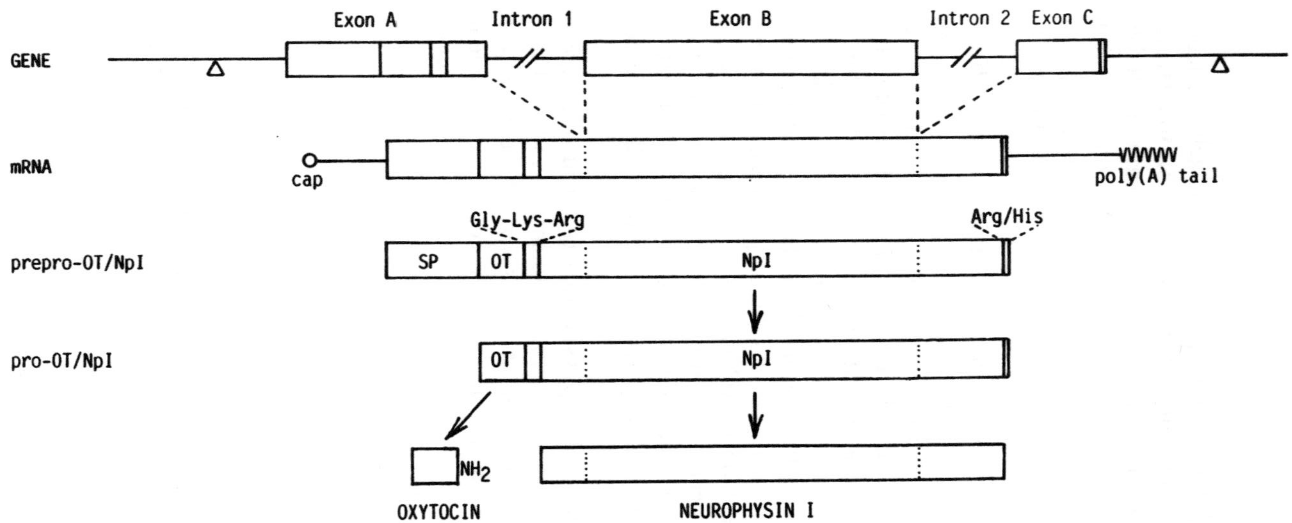

Fig. 3. Scheme representing the biosynthesis of oxytocin and neurophysin I based on data obtained from the rat and cow. See text for details

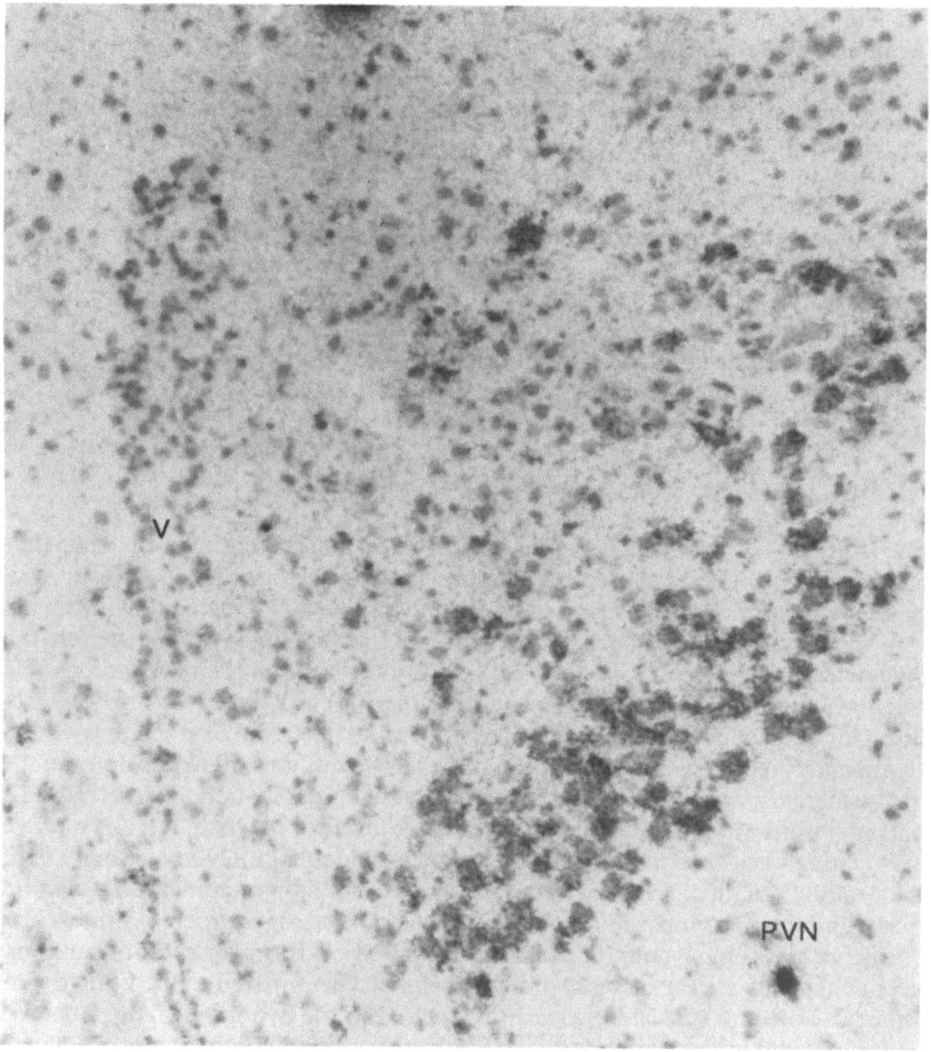

Fig. 4. In situ hybridization of rat paraventricular nucleus (*PVN*) using a 600 bp oxytocin-specific DNA probe labelled with ^{35}S. Silver grains are clearly aggregated over a discrete population of magnocellular neurones. Cell nuclei counterstained with hematoxylin. *V*, third ventricle. (Courtesy of Dr. Susanne Fehr, Hamburg)

tocin gene specific DNA probes, showed an increase in oxytocin mRNA levels in the SON neurones from late pregnant as opposed to nonpregnant rats. With an alternative technique – solution hybridization to RNA extracted from single hypothalamic nuclei – Burbach et al. (personal communication) have reached similar conclusions.

The role of oxytocin in pregnancy and lactation does not explain why in the male rat just as much oxytocin mRNA is present in the hypothalamus as in the female. Neurophysiological studies have recently shown that in the SON the oxytocin neurones react just as readily to osmotic stress as do vasopressin neurones (Brimble and Dyball 1977; Leng et al. 1982). In fact, biosynthesis of pituitary oxytocin has been shown to be increased two- to threefold, equivalent to the change in vasopressin synthesis, in rats given 2% saline instead of water for 5 days prior to experimentation (Franco-Bourland and Fernstrom 1981). Assessment of total hypothalamic mRNA either by in vitro translation and measurement of the oxytocin precursor (Sherman and McKelvy 1983) or by Northern blot analysis with an oligonucleotide probe (Majzoub et al. 1983) suggested that dehydration stimulated the transcription of the oxytocin as well as the vasopressin gene. Using the more sensitive technique of solution hybridization of RNA extracted from individual hypothalamic nuclei, Van Tol et al. (manuscript in preparation) have shown a pronounced increase upon osmotic stress in the specific levels of both oxytocin and vasopressin mRNA in the SON and PVN.

7.1 The Brattleboro Rat as a Model for Oxytocin Synthesis

The Brattleboro strain of rats in characterized by an inherent hypothalamic diabetes insipidus. This is caused by a failure to produce vasopressin, a result of a single base deletion in the vasopressin gene (Schmale and Richter 1984; Schmale et al. 1984). As a consequence, in the hypothalamo-hypophyseal system only the oxytocinergic neurones appear to be functioning normally. Given the possible role oxytocin may be playing in regulating the water balance (see previous section), it is particularly interesting to see what adjustments to oxytocin biosynthesis are made in the Brattleboro rat. Firstly, the afferent noradrenergic input to the oxytocinergic magnocellular neurones is increased (Schöler and Sladek 1981). To date no information is available on the relative levels of oxytocin mRNA in these cells, though cell-free translation of total hypothalamic mRNA and analysis of the products using antineurophysin antisera do not suggest a marked deviation from nonmutant strains (Fig. 5). Calculations of rates of oxytocin synthesis and release in the homozygous Brattleboro rat based on turnover rates of pituitary peptide indicate a clear increase in both parameters compared to the Long Evans strain (North et al. 1982). Release of oxytocin into the hypophyseal portal blood is also elevated in the Brattleboro rat (Horn et al. 1985). Whether these effects can be atttributed to the cross-reactivity of oxytocin with vasopressin receptors and hence a spurious feedback of information to the brain, or whether oxytocin can act independently as an osmoregulator is not clear. The Brattleboro rat should be a useful model to test this hypothesis.

7.2 Ontogeny of the Hypothalamic Oxytocinergic System

Analysis of human fetal material showed oxytocin to be first detectable in the hypothalamus and pituitary at 14 weeks of gestation; the pituitary content thereafter increases by three orders of magnitude over the next 12 weeks (Burford and Robinson 1982). Neurophysin I appears about 2 weeks prior to the nonapeptide, implying that although synthesis of the precursor polyprotein may be occur-

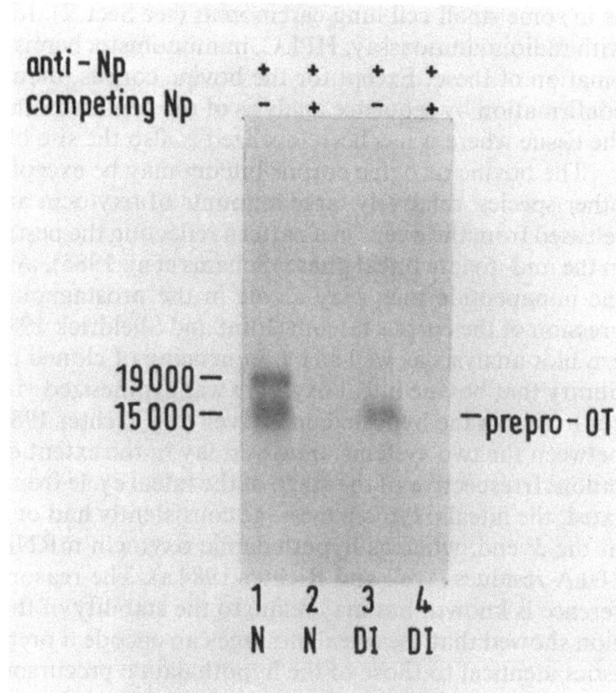

Fig. 5. In vitro translation of hypothalamic mRNA from Wistar (*N;* lanes 1 and 2) or Brattleboro (*DI;* lanes 3 and 4) rats. The translation products were immunoprecipitated with an antineurophysin antiserum which recognizes both oxytocin (M_r 15000) and vasopressin (M_r 19000) precursors. Lanes 2 and 4 are controls where cold mixed bovine neurophysins I and II have been added as competitive antigens during immunoprecipitation. See Schmale et al. (1984) for details

ring at this earlier time, the post-translational processing system is not yet mature.

Solution hybridization of RNA from postnatal rat hypothalamic nuclei showed a moderate rise in the PVN content of oxytocin mRNA between day 1 and day 30, but a much steeper increase in the SON over the same period (Van Tol et al. Manuscript in preparation). Since this latter rise parallels closely both the vasopressin mRNA content of the same nucleus and the onset of kidney function, it has been postulated that the ontogeny of both hormones may be linked to the postnatal development of osmoregulation (Van Tol et al. manuscript in preparation).

8 Regulation of Oxytocin Gene Expression in Peripheral Tissues

Oxytocin and neurophysin I have been detected in several peripheral organs: the placenta, the testis, the adrenal gland, the ovary, and the corpus luteum, as well as in some small cell lung carcinomas (see Sect. 2). Identification has been made with radioimmunoassay, HPLC, immunohistochemistry, bioassay, or some combination of these. Except for the bovine corpus luteum there is neither chemical confirmation by sequence analysis of the peptide or its mRNA, nor evidence that the tissue where it has been localized is also the site of synthesis.

The bovine or ovine corpus luteum may be exceptional in that, in contrast to other species, relatively large amounts of oxytocin are present. This oxytocin is released from the ovary in a pattern reflecting the oestrous cycle, with a maximum in the mid- to late luteal phase (Schams et al. 1985), and it has been suggested that the nonapeptide may play a role in the prostaglandin-F2 alpha controlled regression of the corpus luteum (Flint and Sheldrick 1982). It was shown by Northern blot analysis as well as by sequencing of cloned cDNA from a luteal cDNA library that bovine luteal oxytocin was synthesized via a similar common precursor to that in the hypothalamus (Ivell and Richter 1984a). An essential difference between the two systems, however, lay in the extent of the 3′ mRNA polyadenylation. Irrespective of the stage of the luteal cycle from which the mRNA was isolated, the luteal oxytocin message consistently had only ca. 40 adenosine residues at the 3′ end, whereas hypothalamic oxytocin mRNA had an appendage of ca. 100 A residues (Ivell and Richter 1984a). The reason for this tissue-specific difference is known, but may relate to the stability of the mRNA. Cell-free translation showed that the luteal message can encode a preprooxytocin with characteristics identical to those of the hypothalamic precursor (Ivell and Richter 1984a).

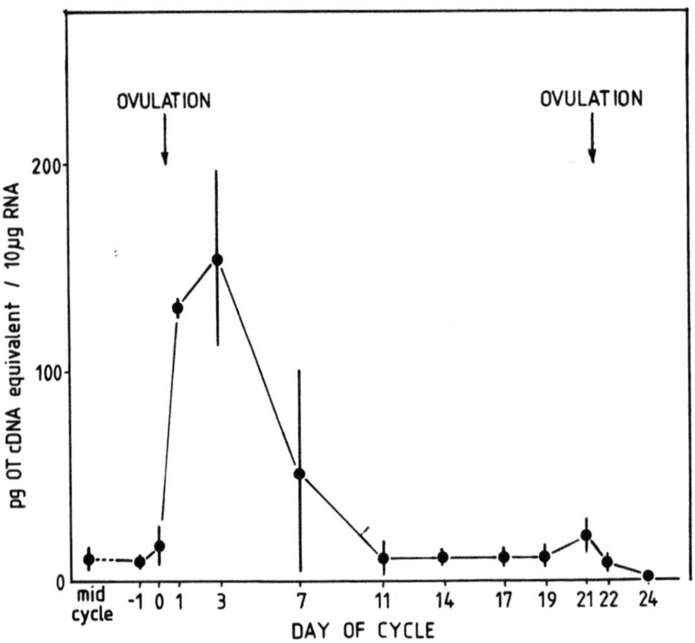

Fig. 6. Oxytocin mRNA levels in the nonpregnant bovine corpus luteum through the oestrous cycle, measured by dot hybridization. See Ivell et al. (1985) for details

In vivo studies have reached a similar conclusion in regard to pro-oxytocin (Swann et al. 1984).

Using a 3′ oxytocin-specific DNA probe it has been possible to map, via dot blot analysis of extracted RNA, the precise time in the luteal cycle when the oxytocin gene is most active (Fig. 6; Ivell et al. 1985). Although oxytocin mRNA is already detectable at a low level in the preovulatory follicle, there is a large surge of oxytocin gene transcription only upon ovulation. This peak of gene activity is short-lived, oxytocin mRNA levels being already considerably reduced by day 7 of the 21-day bovine cycle. Thereafter, levels remain low until complete regression of the corpus luteum after day 24. The corpus luteum during pregnancy contains negligible oxytocin mRNA or oxytocin peptide (Schams et al. 1985; Ivell et al. 1985).

Measurements of oxytocin mRNA levels in the bovine hypothalamus at different points through the oestrous cycle showed no significant trend (Ivell et al. 1985). This implies that the factors regulating transcription of the oxytocin gene are tissue-specific, the hypothalamus being regulated independently of the corpus luteum.

9 Outlook

Two important new developments have taken place in the study of oxytocin endocrinology. The first was the impact of modern molecular biological techniques in elucidating the details of the oxytocin biosynthetic pathway. The second was the recognition that oxytocin probably subserves a very broad range of functions in a classically hormonal, paracrine, or neurotransmitter mode, and is probably synthesized in an equally diverse number of tissue types. The future is going to see the bringing together of these two fronts – the development of very sensitive techniques such as in situ hybridization or solution hybridization, and their application to probe the endocrinology of the "new" functions at the level of oxytocin gene activation. A further new direction will be the investigation of cultured cell systems or other models where the mechanisms involved in oxytocin biosynthesis can be investigated and dissected. Finally, it is to be hoped that the knowledge gained can be applied successfully in the many clinical situations where oxytocin, in the male or the female, is going to be involved.

Acknowledgements. I should like to thank Professor D. Richter and my colleagues at the Institut für Zellbiochemie und klinische Neurobiologie in Hamburg for their help and criticism, and the Deutsche Forschungsgemeinschaft for financial support. I am particularly grateful to Dr. Burbach and Dr. Van Tol of Utrecht University, Dr. Schams of the Technische Hochschule, Munich, and Dr. McCabe and colleagues of the Rockefeller University, New York, for access to their results prior to publication, and to Dr. Susanne Fehr for the inclusion of Fig. 4.

References

Acher R (1979) Neurophysine – molekulare und zelluläre Aspekte. Angew Chemie 91:905–919

Ang VTY, Jenkins JS (1984) Neurohypophyseal hormones in the adrenal medulla. J Clin Endocrinol Metab 58:688–691

Angal S, Chaiken M (1982) Interdependence of neurophysin self-association and neuropeptide hormone binding is expressed by quantitative affinity chromatography. Biochemistry 21:1574–1580

Baltimore D (1981) Gene conversion: some implications for immunoglobulin genes. Cell 24:592–594

Bargmann W, Scharrer E (1951) The site of origin of the hormones of the posterior pituitary. Am Sci 39:255–259

Bradbury AF, Finnie MDA, Smyth DG (1982) Mechanism of C-terminal amide formation by pituitary enzymes. Nature 298:686–688

Breslow E (1979) Chemistry and biology of the neurophysins. Annu Rev Biochem 48:251–274

Brimble MJ, Dyball REJ (1977) Characterization of the responses of oxytocin and vasopressin secretory neurones in the supraoptic nucleus to osmotic stimulation. J Physiol 271:253–271

Burford GD, Robinson ICAF (1982) Oxytocin, vasopressin and neurophysins in the hypothalamo-neurohypophyseal system of the human fetus. J Endocrinol 95:403–408

Camier M, Barre N, Cohen P (1985) Hypothalamic biosynthesis and transport of neurophysins and their precursors to the rat brain stem. Brain Res 334:1–8

Dierickx K, Vandesande F (1979) Immunocytochemical demonstration of separate vasopressin-neurophysin and oxytocin-neurophysin neurons in the human hypothalamus. Cell Tissue Res 196:203–212

Fields PA, Eldridge RK, Fuchs A-R, Roberts RF, Fields MJ (1983) Human placental and bovine corpora luteal oxytocin. Endocrinology 112:1544–1546

Flint APF, Sheldrick EL (1982) Ovarian secretion of oxytocin is stimulated by prostaglandin. Nature 297:587–588

Franco-Bourland RE, Fernstrom JD (1981) In vivo biosynthesis of the L-(^{35}S)Cys-arginine vasopressin, -oxytocin and -somatostatin: rapid estimation using reversed phase high pressure liquid chromatography. Endocrinology 109:1097–1106

Furutani Y, Morimoto Y, Shibahara S, Noda M, Takahashi H, Hirose T, Asai M, Inayama S, Hayashida H, Miyata T, Numa S (1983) Cloning and sequence analysis of cDNA for ovine corticotropin releasing factor precursor. Nature 301:537–540

Gainer H, Sarne Y, Brownstein M (1977) Biosynthesis and axonal transport of rat neurohypophyseal proteins and peptides. J Cell Biol 73:366–381

George JM (1978) Immunoreactive vasopressin and oxytocin: concentration in individual human hypothalamic nuclei. Science 200:342–343

Guldenaar SEF, Pickering BT (1985) Immunocytochemical evidence for the presence of oxytocin in rat testis. Cell Tissue Res 240:485–487

Horn AM, Robinson ICAF, Fink G (1985) Oxytocin and vasopressin in rat hypophyseal portal blood: experimental studies in normal and Brattleboro rats. J Endocrinol 104:211–224

Ivell R, Richter D (1984a) The gene for the hypothalamic peptide hormone oxytocin is highly expressed in the bovine corpus luteum: biosynthesis, structure and sequence analysis. EMBO J 3:2351–2354

Ivell R, Richter D (1984b) Structure and comparison of the oxytocin and vasopressin genes from rat. Proc Natl Acad Sci USA 81:2006–2010

Ivell R, Richter D (1985) The oxytocin gene and its expression in the hypothalamus and ovary. In: Amico JA, Robinson AG (eds) Oxytocin, clinical and laboratory studies. Elsevier Biomedical, Amsterdam, pp 115–123

Ivell R, Brackett K, Fields MJ, Richter D (1985) Ovulation triggers oxytocin gene expression in the bovine ovary. FEBS Lett 190:263–267

Land H, Schütz G, Schmale H, Richter D (1982) Nucleotide sequence of cloned cDNA encoding bovine arginine vasopressin-neurophysin II precursor. Nature 295:299–303
Land H, Grez M, Ruppert S, Schmale H, Rehbein M, Richter D, Schütz G (1983) Deduced amino acid sequence from the bovine oxytocin-neurophysin I precursor cDNA. Nature 302:342–344
Leng G, Mason WT, Dyer R (1982) The supraoptic nucleus as an osmoreceptor. Neuroendocrinology 34:75–82
Majzoub JA, Rich A, van Boom J, Habener JF (1983) Vasopressin and oxytocin mRNA regulation in the rat assessed by hybridization with synthetic oligonucleotides. J Biol Chem 258:14061–14064
Makino T, Nakazawa K, Ishii K, Haginiwa I, Nakayama A, Iizuka R (1983) Detection of immunoreactive human placental oxytocin and its contractile effect on the uterine muscle. Endocrinol Jpn 30:389–395
Martin R, Voigt KH (1981) Enkephalins co-exist with oxytocin and vasopressin in nerve terminals of rat neurohypophysis. Nature 289:502–504
Maurer LH, O'Donnell JF, Kennedy S, Faulkner CS, Rist K, North WG (1983) Human neurophysins in carcinoma of the lung: relation to histology, disease stage, response rate, survival, and syndrome of inappropriate antidiuretic hormone secretion. Cancer Treat Rep 67:971–976
McCabe JT, Morrell JI, Richter D, Pfaff DW (1985) Localization of neuroendocrinologically relevant RNA in brain by in situ hybridization. Frontiers in neuroendocrinology 86
McKelvy JF (1975) Phosphorylation of neurosecretory granules by cAMP-stimulated protein kinase and its implications for transport and release of neurophysin proteins. Ann NY Acad Sci 248:80–91
Moore SEH, Hope DB (1985) High pressure liquid chromatography of bovine neurophysin isoforms. British Neuroendocrine Group inaugural meeting, 10–11 July 1985, Oxford, England, abstract 32
Nicholson HD, Swann RW, Burford GD, Wathes DC, Porter DG, Pickering BT (1984) Identification of oxytocin and vasopressin in the testis and in adrenal tissue. Regul Pept 8:141–146
North WG, Gellai M, Hardy G (1982) Oxytocin and oxytocin-associated neurophysin evaluated by RIA in the Brattleboro rat: turnover. Ann NY Acad Sci 394:167–172
Richter D (1983) Synthesis, processing and gene structure of vasopressin and oxytocin. Prog Nucl Acids Res Mol Biol 30:245–266
Ruppert S, Scherer G, Schütz G (1984) The primary structure of the bovine genes encoding the vasopressin and oxytocin precursor proteins reveals a recent gene conversion event. Nature 308:554–557
Russell JT, Brownstein MJ, Gainer H (1980) Biosynthesis of vasopressin, oxytocin and neurophysins: isolation and characterization of two common precursors (propressophysin and pro-oxyphysin). Endocrinology 107:1880–1891
Sachs H, Fawcett P, Takabatake Y, Portanova R (1969) Biosynthesis and release of vasopressin and neurophysin. Rec Progr Horm Res 25:447–491
Schams D, Schallenberger E, Meyer HHD, Bullermann B, Breitinger H-J, Enzenhofer G, Koll R, Kruip TAM, Walters DL, Karg H (1985) Ovarian oxytocin during the estrous cycle in cattle. In: Amico JA, Robinson AG (eds) Oxytocin, clinical and laboratory studies. Elsevier Biomedical, Amsterdam, pp 317–334
Schmale H, Richter D (1980) In vitro biosynthesis and processing of composite common precursors containing amino acid sequences identified immunologically as neurophysin I/oxytocin and as neurophysin II/vasopressin. FEBS Lett 121:358–362
Schmale H, Richter D (1984) Single base deletion in the vasopressin gene is the cause of diabetes insipidus in Brattleboro rats. Nature 308:705–709
Schmale H, Leipold B, Richter D (1979) Cell-free translation of bovine hypothalamic mRNA: synthesis and processing of the prepro-neurophysin I and II. FEBS Lett 108:311–316
Schmale H, Ivell R, Breindl M, Darmer D, Richter D (1984) The mutant vasopressin gene from diabetes insipidus (Brattleboro) rats is transcribed but the message is not efficiently translated. EMBO J 3:3289–3293

Schöler J, Sladek JR (1981) Supraoptic nucleus of the Brattleboro rat has an altered afferent noradrenergic input. Science 214:347–349

Sherman TG, McKelvy JF (1983) Cell-free biosynthesis of rat neurophysin polypeptides from poly(A)RNA isolated from individual hypothalamic nuclei. Neurosci Abstr 9:622

Sofroniev MV (1983) Morphology of vasopressin and oxytocin neurones and their central and vascular projections. In: The neurohypophysis: structure, function and control. Prog Brain Res 60:101–114

Sofroniev MV (1985) Vasopressin- and neurophysin-immunoreactive neurones in the septal region, medial amygdala and locus coeruleus in colchicine-treated rats. Neuroscience 15:347–358

Swann RW, Gonzalez CB, Birkett SD, Pickering BT (1982) Precursors in the biosynthesis of vasopressin and oxytocin in the rat. Biochem J 208:339–349

Swann RW, O'Shaughnessy PJ, Birkett SD, Wathes DC, Porter DG, Pickering BT (1984) Biosynthesis of oxytocin in the corpus luteum. FEBS Lett 174:262–266

Van Dyke HB, Chow BF, Greep RO, Rothen A (1942) Isolation of a protein from the ox pituitary with constant oxytocic, pressor and diuresis-inhibiting activities. J Pharmacol Exp Ther 74:190–209

Wathes DC, Swann RW (1982) Is oxytocin an ovarian hormone? Nature 297:225–227

Whitnall MH, Gainer H, Cox BM, Molineux CJ (1983) Dynorphin-A-(1-8) is contained within vasopressin neurosecretory vesicles in rat pituitary. Science 205:1137–1139

Wolfson B, Manning RW, Davis L, Arentzen R, Baldino F (1985) Co-localization of corticotropin releasing factor and vasopressin mRNA in neurons after adrenalectomy. Nature 315:59–61

Regulation of Oxytocin Release

M. L. FORSLING

Contents

1 Introduction

The first described, and still the most important, actions of oxytocin are its contractile effects on uterine smooth muscle (Dale 1906) and myoepithelial cells (Ott and Scott 1910). Not unnaturally, therefore, early investigators sought to establish release of the hormone during labor and lactation and found that this was indeed the case (see Tindal 1975). The highest concentrations of the hormone in plasma have been found at these times. Release in the female has also been claimed at other times of the reproductive cycle, including mating. Stimuli for secretion in the male animal are not so well established. Again it has been suggested that release also occurs during mating and that it may be associated with known stimuli for vasopressin, such as increased plasma osmolality and hemorrhage (Forsling and Brimble 1985). As with most hormones, understanding of the

Department of Physiology, The Middlesex Hospital Medical School, Cleveland Street, London W1P 6DB, Great Britain mechanisms of release has depended on the availability of sensitive assay techniques. Sensitive bioassays have been available for a number of years (Chard and Forsling 1976), but the number of samples which can be processed is very limited. Release has been monitored in many studies by recording intramammary pressure, but of course this approach is limited to the lactating female. Recordings have been made from putative oxytocin neurons (see chapter by Dreifuss), but again this approach is largely limited to studies in the lactating female. A sensitive radioimmunoassay for oxytocin was described in 1969 and was subsequently applied to many investigations (Chard and Forsling 1976), but it is only relatively recently that sensitive radioimmunoassays have been established in a number of centers.

For many years it was believed that oxytocin was synthesized only in the magnocellular neurons of the supraoptic and paraventricular nuclei and was stored in the posterior pituitary. It has now been demonstrated that fibers containing oxytocin are found in many regions of the brain (Sterba 1979), and it has been suggested that oxytocin is released from these fibers into the cerebrospinal fluid (CSF) independently of its secretion into the systemic circulation. The presence of oxytocin has also been demonstrated in the ovaries (Wathes 1984) and the testes and adrenal tissue (Nicholson et al. 1984) of certain species. This review will only touch on these aspects of oxytocin, which will be described in more detail in the chapters by Wathes and Burbach (this volume). This chapter will deal with oxytocin release over the reproductive cycle, certain aspects of which will be amplified in the chapter by Robinson (this volume), and other factors causing oxytocin release, with observations on the Brattleboro rat. In this animal there is a congenital absence of vasopressin, and it has been suggested that oxytocin, since some of its actions overlap with those of vasopressin and since neurons containing each hormone are found close together in the hypothalamus, might subserve some vasopressin function. Mention will be made of the central control of hormone release, but this has been covered more fully in the chapter by Dreifuss.

A wide range of values have been given for the basal oxytocin concentration in male and female subjects. The magnitude of the estimate of the hormone concentration in the plasma probably depends on the nature of the antiserum employed by each individual worker. The results of radioimmunoassay for vasopressin have similarly been shown to depend on the particular antiserum employed (Forsling 1976; Thomas and Lee 1976). However, the general consensus appears to be that the circulating concentration of oxytocin is similar to that of vasopressin, of the order of 1 μu/ml or 2 pg/ml.

There are relatively few data as to whether oxytocin concentrations vary with age, but there is evidence from studies of the rat that sexual maturation plays a role in increasing the synthesis of oxytocin in the hypothalamic nuclei (Jackson and George 1980). There is also evidence from the rat of increased oxytocin secretion with advancing age (Fliers and Swaab 1983). Landgraf et al. (1982a) found evidence of a circadian variation in plasma oxytocin concentrations in man which decreased from 8:00 A.M. to 4:00 P.M. However, Amico et al. (1983), although able to demonstrate a time-dependent peak in oxytocin at around 12:00 A.M. in human CSF could find no evidence of a diurnal variation in oxytocin and its associated neurophysin in plasma. Indeed, the alterations in CSF oxytocin in primates in general do not appear to be reflected in plasma concentrations (Perlow et al. 1982), which suggests that the mechanisms controlling release into the CSF may be different from those influencing release into plasma. There is also a difference between species in the secretion of oxytocin into the CSF. Whereas a diurnal rhythm is seen in primates, no such pattern is seen in the cat or rat (Reppert et al. 1983).

Considerable differences between species are also seen in the release of oxytocin during parturition and lactation. Unfortunately, this has led to some confusion when investigators have attempted to extrapolate from the results in one animal species to another. Such variation among species is hardly surprising when one considers the different requirements of various mammals. There are obviously quite different problems and solutions for the delivery, for example, by a biped of a single infant weighing 6%–7% of maternal body weight and the ordered delivery by a quadruped, such as the pig, of many offspring, each weighing about 0.5% of maternal body weight. Equally, in the cow and the goat the cervix is gradually dilated, while in the rabbit, where the birth canal is long in comparison to the length of the umbilical cord, delivery must be rapid to prevent suffocation of the young. Similarly, in lactation there are many patterns of behavior, depending inter alia on the maturity of the young. In the rat, where the young are relatively immature, suckling occurs throughout the day, whereas the more mature guinea pig young suckle once daily. There are also considerable differences in the mammary gland. Whereas in ruminants milk is stored in large cisterns, storage capacity in animals such as pigs is relatively limited.

2 Oxytocin Release in the Fetus and Neonate

Oxytocin was demonstrated in the fetal sheep pituitary in 1937 (Bell and Robson 1937), an observation subsequently confirmed in other species (Dicker and Tyler 1953). Oxytocin has subsequently been demonstrated in the fetal pituitary of many species (Forsling 1977), but it has been suggested that some of the oxytocic activity in the fetal sheep pituitary and possibly that of other species is due to the presence of arginine vasotocin (Vizsolyi and Perks 1976; Skowsky and Fisher 1977). The presence of neurosecretory material in the fetal hypothalamus was established histochemically by Benirschke and Mackay (1953). Although Alexander et al. (1973) found little neurosecretory material in sheep fetuses of 80 days' gestation (full term 147 days), nevertheless oxytocin was present in considerable quantities, and the amount increased with increasing gestational age, but not to the same extent as vasopressin.

Higher levels of vasopressin have also been found in the pituitary of the fetal guinea pig (Burton and Forsling 1972), the neurohypophysial system of which has been studied in some detail (Silverman 1975), and it has been suggested that the increase in hormone content before birth is due to the production of a stimulatory factor in the fetal hypothalamo-hypophysial system (Pearson et al. 1975).

Recently the oxytocin content of the human fetal pituitary gland has been investigated (Burford and Robinson 1982), and the hormone has been shown to be present at 14–17 weeks at a level of 10.2 ± 5.9 ng/gland and to increase during gestation to 544.3 ± 33.8 ng/gland in 1- to 5-day-old infants (Khan-Dawood and Dawood 1984).

The increase in the oxytocin content of the pituitary gland with increasing gestational age was due to a significant increase in the oxytocin concentration, rather than to the increase in the weight of the pituitary gland. Since the findings were from glands removed at death or abortion, they may not represent the true oxytocin content under physiological conditions, but do support the hypothesis that the fetal neurohypophysial system is active at birth. Although Forsling (as cited in Fuchs 1973) did not find that the change in the oxytocin content of the fetal rat pituitary at birth was significant, Schriefer and Miller (1979) found a significant fall in the levels in the fetal pituitary gland on spontaneous delivery.

A fall in the pituitary gland oxytocin content on delivery would be consistent with the observed increase in fetal plasma oxytocin concentrations at delivery. Chard et al. (1971) first reported elevated concentrations of oxytocin in cord blood, the levels being higher in the umbilical artery than in the umbilical vein, indicating that the hormone originated from the fetal pituitary. These observations were subsequently confirmed (Kumaresan et al. 1975; Dawood et al. 1978c). Dawood et al. (1978c) reported that the umbilical oxytocin level was higher in women experiencing contractions, whether delivered vaginally or by Cesarian section, than in women without contractions delivered by elective Cesarian section. This suggests that fetal oxytocin may be involved in the initiation of parturition. A recent report that oxytocin concentrations were undetectable in plasma taken in the second stage of labor from an anencephalic fetus when the maternal concentrations were 13.8 ± 8.0 μu/ml confirms that the oxytocin is of fetal rather than maternal origin (Otsuki et al. 1983). The presence of oxytocin in the fetal urine and amniotic fluid also points to the same conclusion (Dawood et al. 1978c).

Oxytocin concentrations in fetal plasma have been followed in more detail in experimental animal preparations. During pregnancy fetal oxytocin concentrations are low or undetectable in the sheep and pig (Alexander et al. 1974; Forsling et al. 1975; Macdonald et al. 1979). However, during late pregnancy fetal plasma oxytocin concentrations have been reported to be higher than the maternal plasma oxytocin concentrations in fetal sheep (Forsling et al. 1975; Mitchell et al. 1982) and baboon (Dawood et al. 1979). Glatz et al. (1981), also studying the sheep, noted a similar trend, but the difference was not statistically significant. No such trend was seen in the fetal pig (Forsling et al. 1979a). Increased plasma oxytocin concentrations at delivery are consistently seen in studies in the sheep, though Glatz et al. (1981) and Mitchell et al. (1980) failed to confirm the previously described increase immediately prior to parturition (Forsling et al. 1975). However, Mitchell et al. took more frequent samples and demonstrated considerable variability in oxytocin concentrations between and within animals, which could account for the differences as Forsling et al. took a very limited number of samples.

In the rat pup after birth it was found that the pituitary levels of oxytocin were low up to 8 days and then increased between 8 and 21 days (Sinding et al. 1980). There are few data on plasma concentrations at this time, but Glatz et al. (1981)

reported that the highest concentrations seen 15 min post partum decrease by 180 min to 8.9±3.2 μu/ml, a value still considerably higher than the baseline figure of 2.7±1.1 μu/ml. Leake et al. (1981 a) also reported a rapid fall in the concentrations by 30 min after delivery of the human infant. However, the plasma oxytocin concentrations remained elevated above basal adult levels (1.7±0.3 μu/ml) during the study, which continued for 4 days post partum. Thus, stimulation of fetal oxytocin release may continue beyond the period of birth.

The stimulus for the release of oxytocin in the fetus during and after labor is obscure. It is not clear whether it is a primary phenomenon or one secondary to the process of labor, nor is it clear whether the release plays a role in parturition. Alexander et al. (1976) failed to obtain oxytocin release in the acutely exteriorized sheep fetus, except in response to an infusion of calcium chloride into the carotid artery, and even then the response was inconsistent. Similarly, Stegner et al. (1984) found that fetal plasma oxytocin levels were unaffected by hypoxemia. Furthermore, in the fetal pig Forsling et al. (1979 b) found that the oxytocin concentrations were unaffected by hemorrhage, which is a potent stimulus of vasopressin release. It is possible that release is due to fetal stress, for, as will be discussed later (Sect. 8), there is recent evidence that oxytocin is released in response to certain stresses. It is also possible that activation of the fetal posterior pituitary in labor and delivery is the result of an acute stimulus provided by uterine contractions to a maturing hypothalamo-neurohypophysial system. Prostaglandins may play a part (Forsling et al. 1979 b). The levels are higher in the fetus than in the mother, but there is no difference between those found at vaginal delivery and at Cesarian section (Craft et al. 1973). Recently it has been reported that dopamine acts on the fetal neuroendocrine system to inhibit oxytocin release (Schriefer et al. 1980). Reduction of this inhibitory input might also be important at the time of parturition.

3 Oxytocin Release During Estrous and Menstrual Cycles

The problems associated with the determination of oxytocin in plasma meant that as with other aspects of posterior pituitary function, the evidence that oxytocin varied over the estrous cycle came with observations on the pituitary content. Heller (1959) reported that the oxytocin content of the rat posterior lobe varies over the estrous cycle, the levels being lower at metestrus. The fact that the neurohypophysial hormone content was linked to the reproductive cycle was subsequently confirmed by Fendler (1961) and Konig and Bottcher (1966) using bioassay, and later by Crowley et al. (1978) employing a radioimmunoassay. A change in the synthetic activity of the supraoptic and paraventricular nuclei has also been reported at different stages of the cycle (Swaab and Jongkind 1970). Basal and potassium-stimulated oxytocin release in vitro have been shown to vary according to the day of the cycle, both being higher during estrous than diestrous (Pitzel et al. 1981). Most evidence suggests that the pituitary content is influenced by estrogen and progesterone (Konig and Ehlers 1973; Piechowiak et al. 1976), the content of oxytocin and vasopressin in ovariectomized rats being increased by estrogen treatment (Heller 1959; Crowley et al. 1978; Skowsky et al. 1979). Swaab and Jongkind (1971), however, argued that the changes were only indirectly influenced by steroids, the gonadotrophins being the factors which influence the activity of the magnocellular neurons.

Plasma concentrations of oxytocin have been measured over the estrous cycle of a number of species, including the sheep, cow, and goat. Lower concentrations are found in the cow than in the goat and sheep (Schams 1983), but overall a similar pattern of secretion is seen. The concentrations of oxytocin in plasma decrease at the time of luteal regression to reach low levels during the preovulatory period and then increase after the preovulatory surge, when the progesterone concentrations are high (Webb et al. 1981; Sheldrick and Flint 1981; Schams et al. 1982b; Homeida and Cooke 1983). As will later be discussed in more detail, ovarian steroids can modify the release of oxytocin from the posterior pituitary, but since the corpus luteum of both cow and sheep contains large quantities of oxytocin (Wathes and Swann 1982) it seems likely that the major contribution of circulating oxytocin comes from the ovary. Burns et al. (1981) have reported briefly on the changes in oxytocin concentrations during the estrous cycle in the mare. In this species, the oxytocin concentrations are lower on days 10–15, which correspond to the mid-luteal phase, than on day 2 of estrous and 5 days after ovulation. This pattern corresponds more closely to that seen in primates. Such differing patterns may reflect the dominant source of the hormone, pituitary or ovary. Control of release from the two sites would be expected to be different.

The first suggestion that neurohypophysial function in primates might vary over the menstrual cycle came with observations on neurophysin. There is evidence to show that release of oxytocin is associated with release of neurophysin I (estrogen-stimulated neurophysin; Robinson et al. 1977). A mid-cycle increase in this particular neurophysin was seen in the rhesus monkey and was closely correlated with the mid-cycle surge of LH and elevation of estrogen (Robinson et al. 1976). Legros et al. (1975) also detected an elevation of neurophysin mid-cycle in menstruating women. That oxytocin secretion was also elevated was subsequently confirmed by direct estimation of the hormone. Falconer et al. (1980) demonstrated in a group of conscious monkeys that oxytocin concentrations were higher mid-cycle on days 10–11 than on days 4–5 and 28–29. They found no significant correlation between oxytocin and estrogen or progesterone. However, given the differences in the time course of the action and clearance of steroid and small peptide hormones, this does not necessarily rule out a modulating role of steroids in oxytocin release.

An increase in circulating levels of oxytocin at the time of peak levels of LH was reported by Mitchell et al. (1980a) in two women. Amico et al. (1981) confirmed this observation in a more detailed study of five women, in which they measured oxytocin and its associated neurophysin. Kumaresan et al. (1983) also found that oxytocin concentrations were highest at the time when estrogen concentrations would be elevated, and in addition found evidence of a second minor increase during the luteal phase, which was also seen in the study of Mitchell et al. (1980a). The physiological significance of these changes is not clear, but they suggest a possible role of oxytocin at ovulation or in the regulation of luteolysis. Interestingly, Leake et al. (1984) have shown a variation in the oxytocin secretory response to breast stimulation during the menstrual cycle. However, they found that stimulation with a mechanical pump only produced significant increases during the luteal phase at a time when lower oxytocin concentrations have generally been reported. Oxytocin has been found in the human ovary (Khan-Dawood and Dawood 1983; Wathes 1984), but it is unlikely to make a contribution to the ovulatory rise in oxytocin as the concentrations in these studies were too low and the timing was not appropriate.

4 Oxytocin Release During Mating

4.1 The Female

The pattern of oxytocin release over the estrous cycle might be significant with regard to the changes seen during mating. It was first proposed by Harris (1947) that oxytocin might be released during mating and could in turn augment uterine motility and accelerate the movement of spermatozoa. Much indirect evidence was published in support of this hypothesis (see Fitzpatrick 1966). It has been known by herdsmen for centuries that vaginal distension results in milk ejection, and this was confirmed in the laboratory (Andersson 1951; Van Demark and Hays 1953). Further confirmation was provided by bioassay and radioimmunoassay estimates of oxytocin (Roberts and Share 1968; Flint et al. 1975). In addition, mechanical stimulation of the vulva and/or vagina could also produce milk ejection (Van Demark and Hays 1953; Debackere and Peeters 1960).

Although milk ejection has been reported to occur in the human (Harris and Pickles 1953) and in cattle (Van Demark and Hays 1953) during mating, estimates by bioassay have yielded an inconclusive picture on oxytocin release (mares, Walmsley 1963; women, Fox and Knaggs 1969; goats, McNeilly and Folley 1970; rabbits, Sharma and Chaudhury 1970). An increase in plasma concentrations was seen in only a few of the animals studied. More recently Fuchs et al. (1981) employed a radioimmunoassay to investigate the problem. They found that mating produced a rapid transient increase in plasma oxytocin in the rabbit. However, the concentrations achieved were not sufficient to account for the increased uterine activity which was observed. In contrast, there was no release of oxytocin during mating in sheep (Garcia-Villar et al. 1985) nor in cows (Schams et al. 1982a), only massive vaginal distension was effective. It should be pointed out, however, that in an earlier study in the goat there was no evidence that intromission specifically stimulated oxytocin release, which could be brought about for example by the male nuzzling the female (McNeilly and Ducker 1972). Studies in the goat have indicated that auditory, visual and olfactory stimuli are all capable of triggering oxytocin release (McNeilly and Ducker 1972).

4.2 The Male

One proposed role for oxytocin in the male has been in the emission and expulsion of semen, but, as in the female, the results of experiments on the effect of mating on oxytocin release are conflicting. An early experiment by Debackere et al. (1961), in which blood was cross-circulated between a lactating ewe and a ram, revealed that massaging the seminal vesicles of the ram caused a rise in intramammary pressure in the ewe.

These findings were confirmed by direct estimates by bioassay of circulating oxytocin in the bull (Van Dongen 1965; Sharma and Hays 1968) and the ram (Houvenaghel et al. 1967; Sharma et al. 1972). However, Bereznev (1963) and Sybesma (1968) found no enhancement of oxytocin release in the bull. Early radioimmunoassay estimates appeared consistent with the proposition (Sharma et al. 1972), as did the later work of Kumaresan et al. (1979). Two recent studies employing sensitivity-specific immunoassays have again cast doubt on the importance of oxytocin release in the bull during mating. Neither Schams et al. (1982 a) nor Peeters et al. (1983) could detect oxytocin release, which was seen only in response to stimulation with an electroejaculator, or massage of the seminal vesicles and ampullae per rectum. Release in this instance would seem to be due to stimulation of afferent pelvic nerves in the rectal wall. The lack of response in cattle is perhaps not surprising, since mating in this species is completed within a few seconds and with only weak stimulation of the genital tract. This situation is consistent with the view of Cross and Glover (1958) that the ejaculatory reflex is controlled only by sympathetic discharge from the hypothalamus.

In conclusion, the pattern of mating behavior varies from species to species, so it is to be expected that the pattern of hormonal release will also vary.

5 Oxytocin Release During Pregnancy and Parturition

From the discussion of responses during mating, one might predict that oxytocin release in the female would be associated with parturition. This has uniformly been found to be the case. The oxytocin responses during pregnancy are not so clear-cut and have been the subject of debate. In general, oxytocin concentrations in domestic animals are low or undetectable during gestation (sheep, Forsling et al. 1975, Stegner et al. 1984; mare, Allen et al. 1973; sow, Forsling et al. 1979 b; cow, Schams et al. 1979). The concentrations were also described as low, with no rise around term, in rabbits (Fuchs and Dawood 1980), although the actual values were somewhat higher than those reported in the large domestic animals.

The situation is less clear-cut in the human. Apart from the differences in the assay systems used, there is the additional problem in primates of the presence of oxytocinase in the plasma. Hawker (1961) reported significant oxytocinlike activity in late pregnancy, whereas Chard et al. (1970) found that levels were undetectable even at the commencement of labor. In more recent studies it has been suggested that oxytocin is measurable in the peripheral plasma of women throughout pregnancy and that levels rise gradually with gestational age (Vasicka et al. 1978). Fluctuating levels were reported throughout pregnancy by Dawood et al. (1979), with a sharp peak at 39 weeks' gestation. There was a significant correlation between plasma oxytocin and the week of gestation.

Because of difficulty in obtaining serial samples throughout pregnancy in women, Dawood et al. (1978) and Mitchell et al. (1980b) sought to resolve the problem by estimating plasma concentrations of oxytocin throughout the latter half of pregnancy in the baboon and the rhesus monkey. In the baboon, concentrations of the hormone were reported to increase as term advanced, whereas in the monkey levels were generally low and showed no tendency to rise with advancing gestational age. In agreement with this, Leake et al. (1981b) found that baseline plasma oxytocin concentration in women before labor was not significantly different from that in nonpregnant women.

Before the advent of sensitive assays for oxytocin, there was evidence of oxytocin release during parturition from the common observation that milk ejection occurs during parturition in domestic animals (see Ginsburg 1968). In 1948 a report appeared of milk ejection, synchronous with labor pains during delivery, of a woman lactating from a previous pregnancy (Gunther 1948). Milk ejection responses occurring precisely at the time of delivery were later recorded in the rabbit, the oxytocin release appearing to be a single spurt (Cross 1958). Fuchs (1964), recording uterine activity in the rabbit, came to a similar conclusion. Early attempts to measure oxytocin directly in blood from women in labor yielded extremely high values (Juret et al., as cited by Driesche 1961; Hawker and Robertson 1956). An improved bioassay was applied to the problem by Fitzpatrick (1961), working on sheep, and later applied to horses, cows, and goats (Fitzpatrick and Walmsley 1965; Knaggs 1963; Folley and Knaggs 1965), and finally rabbits (Fuchs 1964; Haldar 1970). In general, the patterns of release were very similar. Oxytocin concentrations showed only small changes during labor, but reached a peak at the moment of delivery (second stage).

After parturition, the levels dropped to preparturition values at a rate similar to the rate of disappearance of injected hormone, indicating that secretion ceases abruptly. Bioassay studies also provided further evidence for the postulated spurt release (Haldar 1970). Although there was general agreement as to the pattern of release, there was considerable discrepancy in the values found.

5.1 Parturition in Experimental Animals

Radioimmunoassay studies have tended to confirm the bioassay data (Chard et al. 1970; Allen et al. 1975), but the values reported have been generally lower, although in some studies good correlation between bioassay and radioimmunoassay was reported (Forsling 1970). Radioimmunoassay, of course, allowed more extensive sampling, and a more detailed pattern of secretion has emerged, as shown in Fig. 1, where data from four of seven miniature pigs studied are documented. The litter sizes in these animals ranged from 3–11, and the time between the delivery of the first and the last piglet varied from 87–464 min. During the few hours before delivery, oxytocin concentrations ranged from 0.3–24 μu/ml and, as compared with late pregnancy, plasma progesterone had declined, while estrone and estradiol had increased considerably. Thus, oxytocin release was enhanced during the first stage and could be observed following obvious straining or pressing down by the sow, an observation consistent with the finding of Summerlee

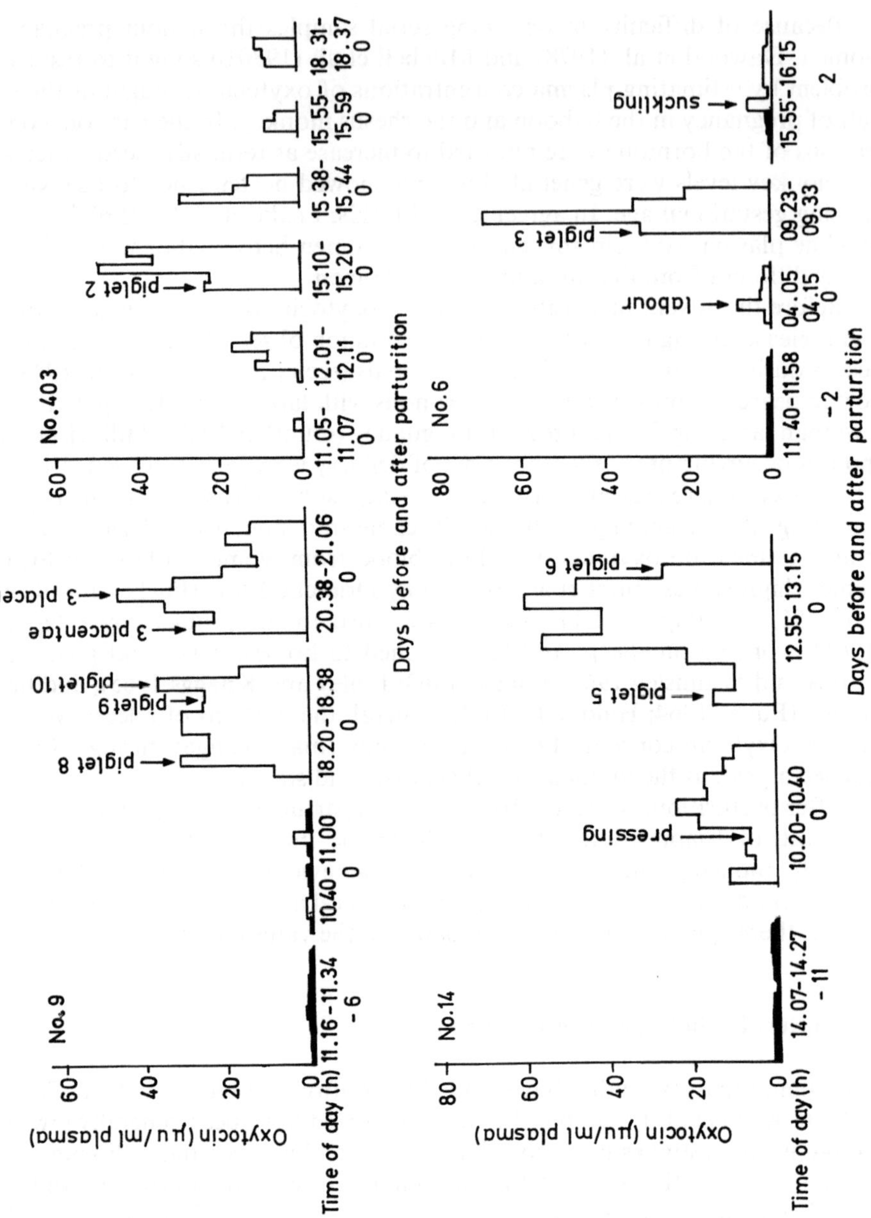

Fig. 1. Oxytocin secretion before, during, and after parturition in four miniature pigs. Delivery of fetuses and placentae is indicated. (Forsling et al. 1979b)

(1983) that abdominal contractions not associated with delivery are followed by activation of oxytocin neurons. Release could be maintained for quite a period of time, as in one animal in which the process of labor was protracted as a result of uterine torsion, oxytocin concentrations remained elevated throughout. After expulsion of the first fetus, the oxytocin concentrations spanned the wide range of 0.8–68.8 μu/ml, peak values on delivery ranging from 13.6–68.8 μu/ml. Expulsion of the placentae was accompanied by an increase in oxytocin concentrations of similar magnitude.

It is noteworthy that in these studies, as in those of Fuchs and Dawood (1980), there was good correlation between oxytocin concentrations and uterine activity (Forsling et al. 1979b; Taverne et al. 1979). This has not always been found to be the case. The basic pattern of activation of the neurohypophysial system during parturition has been confirmed in the rat (Summerlee 1983; Higuchi et al. 1985) and the rabbit (Fuchs and Dawood 1980); in these species, however, delivery sometimes occurred with little or no increase in concentrations of oxytocin, in which case the process was abnormal and protracted.

Changes in clearance of oxytocin do not contribute to the elevated levels, as neither Forsling et al. (1979b) nor Glatz et al. (1981) could find evidence of an altered rate of disappearance of the hormone from plasma at this time. Parturition is an immensely complex process depending on the interplay of a number of factors and those responsible for maternal oxytocin release have not been clearly dissected. The experiments of Forsling et al. (1979b) were not designed to answer the question of whether there was positive feedback on oxytocin release resulting from uterine contractions and further mechanical stimulation of the cervix and vagina. As mentioned, the Fergusson reflex has been confirmed in a number of animal species, and Flint et al. (1978) demonstrated in the sheep near term that vaginal distension produced oxytocin release which could be reversibly blocked by epidural anesthetic, showing that an afferent nervous pathway was involved. In contrast, dilation of the cervix and vagina did not appear to provide a consistent stimulus in the rabbit (Fuchs and Dawood 1980). Further, spinal section at T4–5 or T9–10 did not interfere with oxytocin release.

In addition to neural inputs, oxytocin release could be influenced by a number of humoral factors. The possible effects of prostaglandins have already been mentioned, and there is evidence that circulating concentrations of progesterone and estrogen, as well as regulating the uterine sensitivity to oxytocin, could also modulate its release. Progesterone has been shown to inhibit oxytocin release in the sheep (Roberts and Share 1970; Roberts 1973), and in the pig Forsling et al. (1979b) found that oxytocin release commenced only when progesterone concentrations had fallen to less than 10 ng/ml. The background concentrations of estrogen might be important, but did not seem to be the final uncoupling factor, as the concentration in the pig rose before any increase in oxytocin.

5.2 Parturition in the Human

Despite the theoretical advantages in the use of radioimmunoassay, the application of this technique to studies during human labor has also yielded discrepant results. In the original studies of Chard et al. (1970) no oxytocin was detected in the plasma of women during labor. In later studies, they were able to demonstrate oxytocin in a proportion of samples, which is indicative of spurt release (Gibbens and Chard 1976). The frequency of spurts increased as labor progressed, but maximum concentrations were low and the amplitude of the spikes rarely exceeded 5 μu/ml. On the other hand, the concentrations reported by Kumaresan et al. (1974) and Dawood et al. (1978 b) were elevated at the start of labor and were considerably higher, although spurt release was again noted. The lower reported values would be more consistent with the sensitivity of the uterus as predicted from the magnitude of infused doses necessary for the induction of labor (see Forsling and Chard 1979). The pattern and concentrations observed by Leake et al. (1981 b) were closer to those of Chard et al. The oxytocin concentrations changed little during the first stage of labor, but there was a tendency for the concentrations in postcontraction samples to be higher. The only significant rise was seen with the delivery of the fetus, when plasma concentrations increased from 1.1 ± 0.1 to 4.2 ± 1.1 μu/ml. Although Otsuki et al. (1983) reported an elevation of 12.6 μu/ml at the end of pregnancy, they too found no change in plasma concentrations in the first stage of labor. Spurt release was observed in subjects not in labor, but the frequency of the pulses was less at this time.

While most reports thus indicate oxytocin release in the second stage, even though they disagree as to the levels obtained, there are two publications reporting no significant change in the second stage: Gazarek et al. 1976 and Sellars et al. 1981. Most authors agreed that there was no correlation between uterine activity and plasma oxytocin concentrations, although the mean level of oxytocin during hypocontractile labor has been reported to be low (Amico et al. 1984).

The mechanisms regulating oxytocin release in human parturition are not clear. It would seem that oxytocin is not necessary for the onset of parturition, but that a certain level is required for uterine activity, and that there is an alteration in uterine sensitivity which is a predominant factor in increasing uterine activity. If this is the case, then modulation by gonadal steroids with reduced inhibition could be important. It is generally agreed that there is release of oxytocin during the expulsive phase, the function of which may be to ensure efficient retraction of the uterus once emptied of its contents, and in this instance afferent impulses arriving from the lower birth canal could be involved. Two recent observations would be consistent with this hypothesis. First, it was shown that the process of amniotomy for the induction of labor was associated with spurt release of oxytocin, the higher frequencies being found during sweeping and rupture of the membranes (Chard and Gibbens 1983). Second, while Goodfellow et al. (1983) were able to confirm a significant increase between paired peripheral blood samples taken at the onset of full cervical dilation and crowning of the fetal head in women given pethidine, prochlorperazine mesylate (Stemetil) or nitrous oxide/oxygen as required, they could find no such increase in women with epidural analgesia. Vasicka et al. (1978) had earlier reported attenuation with local block, but not with epidural analgesia.

Another hint at the mechanism of release is the observation that ethanol produces a fall in oxytocin concentrations (Fuchs et al. 1982). Despite the presence of oxytocinase in the plasma, altered metabolism would not seem to influence the concentrations reached, as clearance does not seem to be altered during pregnancy and parturition (Ryden and Sjoholm 1969; Amico et al. 1984).

6 Oxytocin Release During Lactation

That oxytocin was involved in the phenomenon of milk ejection was postulated by Ely and Peterson in 1941. The specificity of the milk ejection response to oxytocin has proved of value in monitoring oxytocin release during lactation. Recording of intramammary pressure allows the timing of release to be followed in the experimental animal and the magnitude of hormone release to be predicted. Furthermore, the response forms the basis of sensitive bioassays.

6.1 In Experimental Animals

The basic pattern of release in lactation, like that in parturition, was originally established by bioassay. Because of the relatively large volumes of plasma required, investigations were carried out on the larger domestic animals – sows, cows, and goats, although some studies were carried out on other species, for example the rabbit (Bisset et al. 1970). It had been found that the effect on the uterus or mamary gland of endogenous release of oxytocin in response to suckling could be simulated by intravenous injection of 100 mu in the sow (Whittlestone 1954), 1000 mu in the goat, and 100–6500 mu in the cow (Donker 1958; Peeters et al. 1960; Chiesa et al. 1967). Overall, the response to the dose in terms of intramammary pressure varies relatively little, so that with large doses there is no increase in the maximum pressure, but multiple contractions develop. This, together with the fact that rapid injection gives an optimal response, hints at spurt release. One would additionally predict that the spurts are not too frequent because of the observed tachyphylaxis. Observations on oxytocin release have tended to be consistent with these predictions.

In the cow, the transient increases in blood levels of oxytocin during milking have been estimated by a number of workers, the concentrations obtained being 15 μu/ml and higher (Hawker 1961; Knaggs 1963; Walmsley 1963; Folley and Knaggs 1966; Pritchard and Hays 1966; Lawson and Graff 1968; Cleverley 1968; Momogen and Schmidt 1970; Randy et al. 1971; Parkash and Andersson 1972). Transient release of oxytocin has also been demonstrated in the ewe during milking and suckling (Fitzpatrick 1961), with concentrations of 3–123 μu/ml being documented. Milking in the goat was as effective a stimulus of oxytocin release as suckling, the concentrations reported being 5–177 μu/ml. Release was seen at any time during suckling in this species and was not necessarily associated with any component stimulus, but rather with the sum of the stimuli. There was considerable variation between individuals and even between different feeds by the same animal. Release in the sow (Folley and Knaggs 1966) was associated with the initial massage by piglets of the mammary gland.

The more recent immunoassay studies have tended to confirm these observations, although, as is frequently the experience, the reported values have in general been lower, with peak concentrations usually below 50 μu/ml (Forsling et al. 1979b; Mosdel et al. 1981; Landgraf et al. 1982b; Robinson and Jones 1982; Gorewit 1983; Fuchs 1984; Schams 1984; Schams et al. 1985). In contrast to some of the earlier reports, oxytocin concentrations seem to be maintained over the period of milking.

One animal in which it is difficult to make direct estimates of plasma oxytocin, but in which nevertheless the neuroendocrine control of milk ejection has been extensively studied, is the rat (Lincoln and Paisley 1982), as milk ejection and the electrical activity of the magnocellular neurons may be conveniently recorded in this species. The rat lies at one extreme of the spectrum of lactation and nursing, in that the young are attached to the nipples for 18 h/day. Milk ejection is about 15 s in duration and is seen at regular intervals of 3–10 min during each suckling period (Wakerley and Lincoln 1971). About 10–12 s before each milk ejection, there is a brief burst of firing of neurons in the supraoptic and paraventricular nuclei of both anesthetized (Lincoln and Wakerley 1974) and unanesthetized rats (Summerlee and Lincoln 1981). However, the response appears specific to oxytocin, in that milk ejection is not associated with increases in circulating vasopressin concentrations (D. W. Lincoln and M. L. Forsling, unpublished observations). Secretion of oxytocin seems to depend on the number of teats stimulated. Milk ejection in the anesthetized rat does not occur when the number of young attached to the nipple is reduced to six or less (Lincoln and Wakerley 1975), and similarly in the rabbit oxytocin levels are much lower when one pup is allowed to suckle, as opposed to the whole litter (Fuchs et al. 1984). Interestingly, in the rat oxytocin release, during suckling occurs only when the EEG is synchronous and comprises large-amplitude slow waves, as seen during sleep (Lincoln et al. 1980). To date the rat appears to be unique in this respect and this, like so many other aspects of lactation, is probably related to the animal's normal environment. In many species suckling could, for example, be a time when the animal is at risk from predators.

The process of lactation has also been studied in some detail in the pig (Forsling et al. 1979b; Ellendorff et al. 1982). The initial investigations in the miniature pig confirmed that oxytocin is released on suckling, the highest concentrations of 2.8–7.4 μu/ml being seen when the piglets vigorously stimulated the mammary gland. When maximum oxytocin concentrations were compared over the course of lactation, it emerged that significantly higher values were caused by suckling in the first 2 weeks than in later stages, a result which has also been reported for the cow (Gorewit et al. 1983), but not the rabbit (Fuchs et al. 1984).

Lactation in the sow is akin to that in the rat, in that many young are suckled at regular intervals during the day and, as with the rat, characteristic patterns of behavior are repeated by both mother and young (Gill and Thompson 1956). Lactation could be initiated by the mother or by the litter (Ellendorff et al. 1982). At first the young were extremely active, nosing at the udders and attaching themselves to the teats. Then followed a well-synchronized phase of quiet suckling, similar to the stretch reaction of rat pups which occurs when the rise in intramammary pressure is maximal (Lincoln et al. 1973). Finally there was a phase of rapid

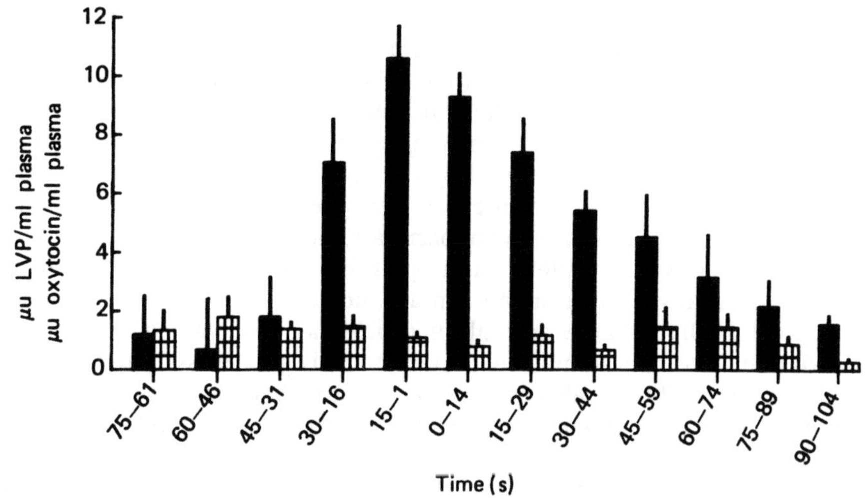

Fig. 2. Plasma oxytocin (*solid bars*) and lysine vasopressin (*LVP, hatched bars*) during milk ejection in the pig. Samples were taken between 75 s before and 104 s after the rise in intramammary pressure, and the results presented as mean ± standard error of mean (SEM). (Ellendorff et al. 1982)

suckling, after which the piglets left the udder. An increase in intramammary pressure occurred 6 min from the onset of suckling and was coincident with the start of the phase of quiet suckling, and it was preceded by some 30 s by an increase in circulating concentrations of oxytocin with peaks of 3.9–21.2 μu/ml (Fig. 2). The initial phase of nursing was characterized in the sow by slow grunting, which increased greatly in frequency at about the time of oxytocin release, i.e., some 23 s before the rise in intramammary pressure. As in the rat and the rabbit (Bisset et al. 1970), no release of vasopressin was observed. Natural milk ejection was best reproduced by rapid injection of oxytocin or by a brief train of electrical stimuli. These observations, together with the described single peak of oxytocin, suggest that oxytocin release is pulsatile.

The pattern of events associated with suckling was very reproducible and occurred at fairly regular intervals of about 45 min. Just as Lincoln (1974) observed that this interval could not be reset in the rat, so if the piglets attempted to suckle early there was "incomplete suckling" (apparently normal suckling which does not lead to milk ejection; Barber et al. 1955), with no oxytocin release. Incomplete suckling was also seen when the mother was disturbed. Although the EEG trace during nursing is always one of wakefulness (Poulain et al. 1981), reflex milk ejection appears to occur only if the sow is in a state of relaxation. When the neurohypophysial system was electrically stimulated, it was found that facilitation of release occurred within a limited range of high-frequency stimuli, which had been previously reported in the rat (Nordman and Dreifuss 1972; Wakerley and Lincoln 1973).

The afferent input for oxytocin release is relatively poorly characterized. Stimulation of the nipple is involved, as are the presence of the young and other factors associated with the routine of nursing or milking. The relative importance of the individual stimuli depends on the individual species. The nipple and skin covering the mammary gland have a rich supply of sensitive nerve endings responding to touch and pressure (Findlay 1966). It is the proprioceptive innervation of the nipple which is involved in reflex milk ejection (Sala et al. 1974). The importance of this afferent input is illustrated by the observation that in the rabbit, which does not appear to release oxytocin in response to sight, sound, or smell of the young, reflex milk ejection does not occur after local anesthetization of the nipple with lignocaine (Findlay 1968). In the goat, however, where oxytocin appears to be released in response to many stimuli, including the sight and sound of the young (McNeilly 1972), a normal milk yield can be obtained from denervated glands (which could also indicate that oxytocin is not necessary for milk removal in this species). There is also evidence in the cow that stimuli associated with the presence of the calf can alter maternal secretion of oxytocin (Akers and Lefcourt 1982).

The mechanism which brings about repeated milk ejection has also to be considered. In the rat, the interval between one milk ejection and the next is remarkably regular and cannot be reset by the introduction of additional milk ejections induced by administration of exogenous oxytocin (Lincoln 1974), or release of endogenous oxytocin by stimulation of the posterior pituitary. During the intervals between milk ejections, even though the pups continue to suckle, the electrical activity of the oxytocin neurons is similar to that observed in the absence of suckling. Thus a mechanism would appear to interrupt the afferent impulses from the stimulated mammary gland before they reach the hypothalamic nuclei. The sudden activation of oxytocin cells would take place only after summation of the afferent impulses and opening of a gate.

The intermittent nature of milk ejection in the pig is partly due to behavioral factors, as the piglets do not suckle continuously. However, a refractory period, lasting much longer than the actual period of suckling, does seem to occur after each milk ejection. Incomplete suckling is observed in apparently unstressed sows when the young suckle earlier than usual (Ellendorff et al. 1982). Tachyphylaxis cannot be the reason why several milk ejections do not occur in one suckling period, as intravenous injections given 1 min after a natural milk ejection caused a response. Prolonged inhibition of release at the level of the neurohypophysis also seems unlikely, as brief electrical stimulation causes prolonged mammary gland responses (Ellendorff et al. 1982). The inhibitory mechanism therefore seems to be located centrally within the afferent arc of the milk ejection reflex, as Cross (1955) originally proposed for the rabbit. It may also be activated by stress, as milk ejection commonly fails in stressed animals.

6.2 In the Human

Nursing mothers quite commonly observe that when an infant is nursing on one breast milk is ejected from the other, an observation which extends far back in time (Folley 1969). Is is generally held that oxytocin release is necessary for the infant to obtain sufficient milk, although this has recently been questioned (Lucas et al. 1980). Because of its assumed role in lactation, oxytocin has been used clinically in breast feeding problems, with the hormone being given intranasally (Newton and Egli 1958; Huntingford 1961).

Intramammary pressure during suckling in the human was recorded by Cobo et al. (1967), who published recordings showing multiple pressure waves with peaks at intervals of 1–2 min. This pattern was most closely reproduced by bolus injections of 20–30 mu given at intervals of about 2 min. Lucas et al. (1980) and McNeilly et al. (1983), taking samples at intervals of 1 min or less, obtained results confirming a pulsatile pattern of release, although Weitzman et al. (1980) and Dawood et al. (1981), sampling at intervals of 2–5 min, were unable to do so. Leake et al. (1984) were able to demonstrate that the oxytocin secretory pattern continues in long-term lactation for the first year post partum although, with stimulation by breast pump, concentrations were higher for lactation at 10–90 days than at 90–180 days. Lucas et al. (1980), in addition, found that the concentrations achieved depended on parity, the oxytocin response being significantly greater in multiparous than in primiparous mothers. There is a marked increase in the tactile sensitivity of the human breast within 1–2 days of delivery (Robinson and Short 1977), which would be consistent with an important role for proprioceptive stimuli applied to the nipple.

Other factors must be involved in oxytocin release in the human, as McNeilly et al. (1983) demonstrated that oxytocin was normally released before the infant started suckling. Nursing mothers are also aware that spontaneous milk ejection may occur independently of suckling, and this has now been formally documented (McNeilly and McNeilly 1978).

7 Osmotic Release of Oxytocin

While the main response of the neurohypophysial system to increased osmotic pressure in the extracellular fluid is release of vasopressin, oxytocin release may also occur. A few years after Verney (1946) published his observations that intracarotid injections of hypertonic saline produced an antidiuresis, Andersson (1951) reported that the same maneuver produced milk ejection. Abrahams and Pickford (1954), recording uterine activity in the dog, also concluded that increased osmotic pressure increased oxytocin release. The work of Sundsten and Sawyer (1961), showing that milk ejection was still produced in rabbits with most of the brain resected, indicated that the response resulted from direct stimulation of the neurohypophysial cells. Activation of the neurohypophysial system to release oxytocin during periods of elevated plasma osmolality was further indicated by the observation that drinking saline depletes the pituitary and hypothalamic stores of oxytocin and vasopressin to the same extent (Jones and Pickering 1969; George 1976; Guzek et al. 1981). More recently, radioimmunoassay studies showed that plasma oxytocin concentrations were elevated after 24-h dehydration of rats or on infusion of hypertonic saline into conscious dogs (Dogeterom et al. 1977; Weitzman et al. 1978).

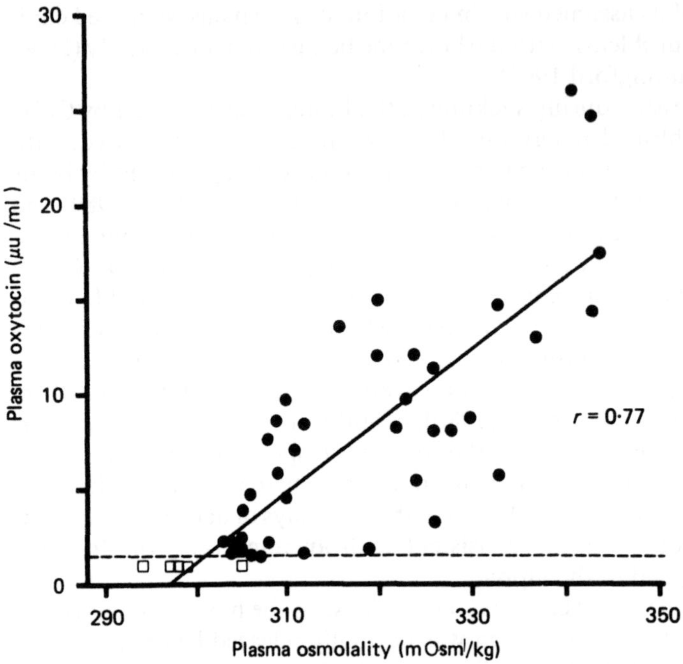

Fig. 3. The relationship of plasma oxytocin to plasma osmolality ($r=0.77$) in conscious rats injected with hypertonic saline (●). Control rats are also shown (□), as is the lower limit of detection (------). (Balment et al. 1980)

Elevated plasma osmolality also increases the firing rate of putative oxytocin neurons in the supraoptic and paraventricular nuclei of the rat (Brimble and Dyball 1977). The relationship between plasma osmolality, firing rate of magnocellular neurons, and plasma oxytocin has now been clearly demonstrated (Brimble et al. 1978). Two injections of hypertonic saline were used to provide a sequential increase in plasma osmolality of approximately 12 mOsm/kg. Fifteen minutes after the first, there was an increase in firing rate from 2.4 ± 0.4 spikes/s to 6.8 ± 0.5 spikes/s with a parallel increase in plasma oxytocin from 2.1 ± 0.3 to 11.6 ± 2.8 μu/ml. After the second injection, the firing rate increased to 9.2 ± 0.5 spikes/s, and the plasma oxytocin concentrations to 14.1 ± 2.7 μu/ml. Similar responses were seen when the plasma osmolality was elevated with hypertonic lithium chloride or mannitol, suggesting that the change in plasma osmolality, rather than plasma sodium, was the stimulus for release.

Elevation of plasma oxytocin in response to elevated plasma osmolality was subsequently demonstrated in the conscious male rat (Balment et al. 1980). The resting plasma osmolality in these animals was 299 ± 2.0 mOsm/kg, but the oxytocin concentrations were undetectable even though the pituitary content was 758 ± 21 mu. Salt loading increased the plasma osmolality to the range 304–344 mOsm/kg (Fig. 3). Oxytocin concentrations were also increased, being linearly related to the osmolality over the whole range. The intercept or osmotic threshold was 297 and the equation of the line is:

plasma oxytocin concentration (μu/ml) = 0.37 × (plasma osmolality − 297 LmOsm/kg).

The regression line is almost parallel to that relating plasma vasopressin to osmolality, though displaced a little to the right of the vasopressin line.

A similar degree of elevation of plasma oxytocin and vasopressin in response to osmotic stimulation has been reported in the unanesthetized dog (Weitzman et al. 1978). An increase in plasma osmolality of 12 mOsm/kg increased plasma oxytocin from 1.2 ± 0.5 to 2.5 ± 0.7 μu/ml, and plasma vasopressin from 1.5 ± 0.2 to 2.4 ± 0.7 μu/ml. Dehydration in the rabbit, too, produced increases in circulating oxytocin of similar magnitude to those for vasopressin (Landgraf and Gunther 1983).

Thus, all the available evidence suggests that the hyperosmolality which accompanies dehydration can stimulate oxytocin release. However, it is not clear whether this is of physiological significance or merely reflects the organization of the neurohypophysial system.

8 Other Stimuli of Oxytocin Release

While the hyperosmolality accompanying dehydration appears to stimulate vasopressin release, the response to the hypovolemic component is not so well established. Using bioassay techniques, Fabian et al. (1979) were able to demonstrate a sixfold rise in plasma oxytocin concentrations on hemorrhage in the rat. The increase, however, was relatively small compared to that of vasopressin. Later Weitzman et al. (1978), using radioimmunoassay, confirmed that a hemorrhage of 12.5 or 25 ml/kg stimulated oxytocin release, but again the stimulation was modest as compared with that for vasopressin. Withdrawal of 20% blood volume in the rat was also seen to cause activation of putative oxytocin neurons (Poulain et al. 1977). However, in the rabbit the concentrations of oxytocin and vasopressin achieved on hemorrhage were of similar magnitude (Landgraf and Gunther 1983).

Since the work of Verney (1947) it has been generally accepted that vasopressin is released in response to stress. The suggestion has recently been made that in the rat oxytocin is also released in response to stress. Lang et al. (1983) recently investigated the effects of exposing rats to three types of stress: physical immobilization for 1 min, swimming in water at room temperature for 1 min, or exposure to cold (4 °C) for 30 min. While exposure to cold had no effect on circulating concentrations of neurohypophysial hormones, immersion in water produced a two- to threefold increase in oxytocin concentrations. Immobilization proved to be a more effective stimulus, producing a five- to tenfold increase in plasma concentrations. Surprisingly, vasopressin release was not stimulated. Gibbs (1984) also found that restraint, but not exposure to cold, stimulated oxytocin release in rats, with no concomitant increase in plasma vasopressin.

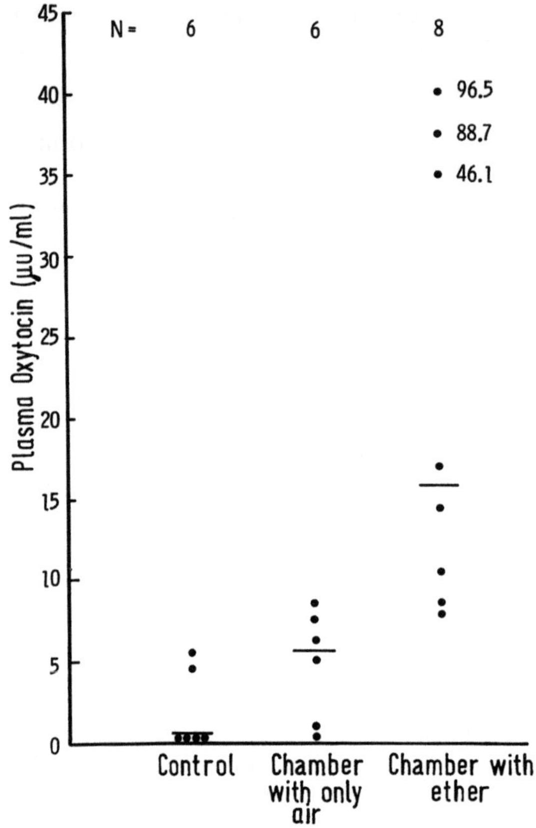

Fig. 4. Plasma oxytocin concentrations in control rats, in a group placed in a chamber containing only air, and in a group exposed to ether vapor for 3 min in the chamber. The median value for each group is presented. (Experiments with Peysner)

Exposure to ether vapor stimulates the release of adrenocorticotropic hormone (ACTH) and is used as the basis of a stress test. Rats have therefore been exposed to ether for 3 min to determine whether this procedure, too, is an effective stimulus of oxytocin release. This proved to be the case, in confirmation of the observations of Gibbs (1984), with very high levels being recorded in some animals (Fig. 4). Inspection of the data also reveals, as had been noted previously (K. Peysner and M. L. Forsling, unpublished observations), that handling of the animals can lead to small increases in basal oxytocin concentrations. However it has been reported in primates that behavioral stress produces a small but significant decrease in oxytocin concentrations (Kalin et al. 1985), an observation which would be consistent with nursing problems observed in women under stress.

9 Oxytocin Release in the Brattleboro Rat

Additional useful insight into the control of oxytocin release can be obtained by the study of the Brattleboro rat with congenital diabetes insipidus (DI). In these animals the turnover of oxytocin is enhanced, possibly to compensate for the lack of vasopressin. The oxytocin cells in the magnocellular nuclei of these animals are hypertrophied (Morris 1982), but administration of vasopressin to normalize urine output and water intake reduces the size to normal. This enhanced production of oxytocin seems to occur even in the fetus, as oxytocin-neurophysin neurons appear earlier in the paraventricular nucleus of fetuses of the homozygous Brattleboro rats than these of Long Evans rats (Boer et al. 1982). Evidence of enhanced oxytocin secretion in the adult was first provided by the observation that the oxytocin content of the pituitary of homozygous DI rats was about 50% lower than in heterozygous or normal animals, and that it could be returned to normal by the administration of vasopressin (Balment et al. 1975; Van Wimersma-Greidanus and de Wied 1977). Concentrations in plasma and CSF were then shown to be higher (Valtin et al. 1965; Dogterom et al. 1977; Balment et al. 1982; Edwards et al. 1982). This could reflect the fact that DI rats are in a fairly continuous state of mild dehydration (Valtin and Schroeder 1964).

Hormone concentrations in these animals increase with increasing plasma osmolality, the line relating these two parameters being virtually superimposable on that obtained for Long Evans rats (Balment et al. 1982). Volume depletion also appears a more effective stimulus for oxytocin release in DI animals than in normal rats (Forsling et al. 1973); this, too, could contribute to any possible oxytocin response to mild dehydration in these animals.

Whether the increase in oxytocin concentrations seen on dehydration in the Brattleboro rat is of functional significance is not clear. Oxytocin has a natriuretic action, but generally in rather higher concentrations than would be produced on dehydration (Balment et al. 1982; Edwards et al. 1982; Robinson et al. 1982). Elevated plasma oxytocin concentrations could, however, account for the shorter gestation and more rapid course of labor in homozygous DI rats (Wong et al. 1982).

10 Central Control of Oxytocin and Putative Neurotransmitters

The central pathways for oxytocin release have been investigated by a number of workers, who have employed the techniques of lesioning, electrical stimulation, and electrical recording. The most systematic studies on the afferent pathways involved in oxytocin release were made by Tindal and Knaggs (Tindal and Knaggs 1971; Tindal et al. 1967; Tindal 1975). However, their results were not always consistent with those of other workers. There have been relatively few studies recently on the pathways involved in milk ejection, and our knowledge is far from complete, but the evidence available indicates that the afferent pathway is via the ipsilateral and lateral fasciculi of the spinal cord (Cross and Dyball 1975). It continues through the lateral tegmentum and divides into three in the rostral midbrain. The most dorsal pathway relays in the medial thalamus and passes to the medial and dorsal regions of the hypothalamus. The intermediate pathway joins the periventricular system, entering the hypothalamus as part of the dorsal longitudinal fasciculus of Schutz. The third pathway enters the ventral hypothalamus as the medial forebrain bundle.

A feature of the scheme presented by Tindal and Knaggs (1975) was a discrete milk ejection pathway in the lateral tegmentum. Other workers have postulated a more diffuse pathway (Richard 1970; Urban et al. 1971). Juss and Wakerley (1981) determined the effects of radio frequency lesions on the milk ejection reflex in the rat and concluded that the lateral tegmentum, which appears to be concerned with transmission from the contralateral nipples, is not an essential component of the reflex, but may contribute to the timing of the intermittent milk ejection responses.

While knowledge of the anatomical pathways involved in oxytocin release is incomplete, much information has accumulated as to the neurotransmitters involved, although this aspect is not free from controversy, either. This may in part be related to the variety of approaches employed. Agonists or antagonists have been administered in a range of doses, intravenously or into the cerebral ventricles of conscious or anesthetized animals, and a variety of responses determined, namely the amounts of milk obtained by or the behavior of the young, milk ejection, the firing patterns of oxytocin neurons, and the circulating concentrations of oxytocin. Acetylcholine, the monoamines, and opioid peptides have all been ascribed a role in oxytocin release.

Acetylcholine, administered either intravenously to the intact animal (Abrahams and Pickford 1954) or topically to the isolated hypothalamus (Bridges et al. 1976), causes the release of oxytocin. Iontophoretic application of acetylcholine to magnocellular neurons results in their excitation, the response being characterized as nicotinic by Dreifuss and Kelly (1972), while Moss et al. (1972) recognized nicotinic and muscarinic responses. More recently Clarke et al. (1978), instead of using agonists, employed cholinoceptor antagonists and investigated their effect on the milk ejection response. Antagonists of the nicotinic type blocked reflex milk ejection. However, they had no effect on the response to the injection of cholinomimetics into the cerebral ventricles, a response which could be blocked by atropine. As most cholinergic synapses in the cerebral cortex are of the muscarinic type (Krnjevic and Phillis 1963), this observation might reflect release of oxytocin in response to influences such as the sight of the young, which would be analyzed by higher centers.

Noradrenalin has been shown to be involved in most types of responses involving oxytocin release, including suckling, vaginal distension, and electrical stimulation of the vagus (Moos and Richard 1975). A twofold mechanism appears to operate – facilitation by the α-receptors and inhibition by the β-receptors. Tribollet et al. (1974) showed that α-adrenergic antagonists inhibited sucking-induced milk ejection, the interval between successive responses being considerably lengthened. On the other hand, Moos and Richard (1979a), who employed specific β-antagonists, were able to confirm an inhibitory role of β-noradrenergic receptors on oxytocin release during suckling. Care has to be taken with the use of systemic noradrenalin, as it can inhibit the response peripherally (Lefcourt and Akers 1984).

Dopamine involvement in oxytocin release has been suggested by the observation that when injected into the third ventricle of anesthetized lactating rats, it produces an increase in intramammary pressure (Bridges et al. 1976). The possibility of dopamine involvement was strengthened by the observation that haloperidol inhibited milk ejection in anesthetized (Clarke and Lincoln 1975) and conscious rats, and had an inhibitory effect on the electrical activity of putative oxytocin cells (Moos and Richard 1982). It was also shown that the potency of dopamine antagonists in blocking reflex milk ejection corresponded closely with comparisons made using dopamine receptor binding (Clarke et al. 1979). More recently, Mason (1983) described excitation by dopamine of putative oxytocinergic neurons in the supraoptic nucleus of the rat in vitro. Investigations have not consistently shown an inhibitory effect, however. Injections of L-dopa were reported to inhibit oxytocin release (Prilusky and Deis 1975). Microiontophoretic application has variously been reported to reduce (Barker et al. 1971) or to enhance (Moss et al. 1972) spontaneous firing.

There are similar discrepancies between the results of in vitro studies. While Bridges et al. (1976) observed stimulation, Seybold et al. (1976) obtained a biphasic inhibitory effect. More recently, Vizi and Volkebas (1980) published evidence indicating that dopamine stored in and released from the neural lobe may inhibit the secretion of oxytocin. They further suggested that the response to opioid peptides of oxytocin release could be effected by removal of the inhibitory effect of dopamine.

There is accumulating histologic evidence that opioid peptides could influence oxytocin secretion, probably at the nerve terminals. Extracts of the neurohypophysis have been shown to contain substantial amounts of Met- and Leu-enkephalin (Kumar et al. 1979), and nerve fibers containing immunoreactive Leu-enkephalin have been described which originate in the magnocellular hypothalamic nuclei and pass to the neural lobe (Rossier et al. 1979). Met-enkephalin appears to be associated with the terminals containing oxytocin immunoreactivity (Martin and Voigt 1981; Vanderhaegenhen et al. 1983).

An inhibitory role of opioid peptides in oxytocin release was revealed by studies in the lactating rat (Haldar and Sawyer 1978), and there is strong evidence that the inhibition is exerted at the level of the posterior lobe. Clarke et al. (1979) found that morphine, while not changing the electrical activity generated by oxytocin cells, reduced the amount of peptide released. A few years later, Bicknell and Leng (1982) were able to demonstrate that naloxone markedly enhances electrically stimulated oxytocin release from the isolated neurohypophysis. The same workers later showed that the potentiating effect of naloxone on hormone release was undiminished after removal of the intermediate lobe tissue so that, consistent with the histologic evidence, the endogenous peptide responsible for oxytocin release appears to come from the neural lobe itself (Bicknell et al. 1983). Reid et al. (1981) were unable to demonstrate any effect on plasma oxytocin concentrations in man after infusion of synthetic β-endorphin, but the basal levels were so low that an inhibitory role could not be excluded.

The neural control of oxytocin is thus seen to be complex. There is evidence of cholingeric, dopaminergic, and opioid pathways, and serotoninergic mechanisms have also been postulated (Mizuno et al. 1967; Moos and Richard 1983). The various pathways could operate at any one of a number of sites. Thus, dopamine and opioid peptides appear to modulate oxytocin release from the posterior pituitary, while dopamine may also act at the supraoptic nucleus cells (Mason 1983), although it is not clear whether it exerts a direct effect or acts on other neurons influencing the supraoptic nucleus. The cholinergic and noradrenergic components may be more distant.

The activity of the neurohypophysial system may also be influenced by the posterior pituitary secretions themselves (Miyake et al. 1983; Freund-Mercier and Richard 1984). Indeed, Freund-Mercier and Richard suggested that oxytocin is one of the main factors governing the periodic simultaneous activation of oxytocin neurons in suckling. Many circulating agents, including active substances released from the ovaries and uterus, are thought to modulate oxytocin secretion. It has been suggested that prostaglandins have a role in release (Gillespie et al. 1972; Ellendorff et al. 1979), possibly in parturition, although Vorherr (1979) could find no evidence that they contribute significantly to oxytocin release in suckling. Relaxin, too, has recently been reported to have a central effect on oxytocin release (Summerlee et al. 1984). However, as indicated in some of the earlier sections, the hormones which have excited most interest in this respect are the ovarian steroids. Roberts and Share (1970) showed that estrogen could enhance and progesterone inhibit oxytocin release in response to vaginal distention. The ovarian steroids have been shown to have an effect on the neurosecretory activity of the hypothalamic neurons (Swaab and Jongkind 1970), and to influence the oxytocin content of the neural lobe (Barnafi and Croxatto 1966; Legros and Grau 1973; Jakubowska-Naziemblo 1979) and the electrical activity of neurons in the paraventricular nucleus, estrogen again being stimulatory and progesterone inhibitory (Negoro et al. 1973). In addition, acute or chronic administration of estrogen to women results in elevation of oxytocin and its associated neurophysin (Amico et al. 1981). That these effects are of physiological significance in some species is indicated by the fact that hormone concentrations vary over the reproductive cycle and that oxytocin release in parturition seems to occur only when progesterone concentrations fall below a certain value (Forsling et al. 1979b).

11 Summary

The recently developed radioimmunoassays have confirmed the results of early bioassay studies on oxytocin release. There are two main occasions on which oxytocin release is significantly augmented: during parturition and lactation, the stimuli for and the importance of release of the hormone varying from species to species. The main stimulatory inputs arise from the vagina or cervix and the nipple. Oxytocin concentrations may also vary with other stages of the reproductive cycle. The stimuli for release in the male are not so well documented, but mating may be one. Plasma concentrations in both sexes may also be influenced by a number of factors seemingly unrelated to reproduction, such as the state of salt and water balance, and stress.

References

Abrahams VC, Pickford M (1954) Simultaneous observations on the rate of urine flow and spontaneous uterine movements in the dog and their relationship to posterior lobe activity. J Physiol (Lond) 126:329–346

Akers RM, Lefcourt AM (1982) Milking and suckling induced secretion of oxytocin and prolactin in parturient dairy cows. Horm Behav 16:87–93

Alexander DP, Britton HG, Forsling ML, Nixon DA, Ratcliffe JG (1973) Adrenocorticotrophin in foetal sheep and the response to stress. In: Pierrepoint C (ed) The endocrinology of pregnancy and parturition: experimental studies in the sheep. Alpha, Omega, Alpha Publishing, Cardiff, pp 112–125

Alexander DP, Britton HG, Forsling ML, Nixon DA, Ratcliffe JG (1974) Pituitary and plasma concentrations of adrenocorticotrophin, growth hormone, vasopressin and oxytocin in foetal and maternal sheep during the latter half of gestation and the response to haemorrhage. Biol Neonate 24:206–219

Alexander DP, Bashore RA, Britton HG, Forsling ML (1976) Antidiuretic hormone and oxytocin release and antidiuretic hormone turnover in the fetus lamb and ewe. Biol Neonate 30:281–287

Allen WE, Chard T, Forsling ML (1973) Peripheral plasma levels of oxytocin and vasopressin in the mare during parturition. J Endocrinol 57:175–176

Amico JA, Seif SM, Robinson AG (1981) Elevation of oxytocin and the oxytocin-associated neurophysin in the plasma of normal women during mid-cycle. J Clin Endocrinol Metab 53:1229–1232

Amico JA, Tenicela R, Johnston J, Robinson AG (1983) A time-dependent peak of oxytocin exists in cerebrospinal fluid, but not plasma of humans. J Clin Endocrinol Metab 57:947–951

Amico JA, Seitchik J, Robinson AG (1984) Studies of oxytocin in plasma of women during hypocontractile labor. J Clin Endocrinol Metab 58:274–279

Andersson B (1951) Some observations on the neuro-humoral regulation of milk ejection. Acta Physiol Scand 23:1–7

Balment RJ, Henderson LW, Oliver JA (1975) The effects of vasopressin on pituitary oxytocin content and plasma renin activity in rats with hypothalamic diabetes insipidus (Brattleboro strain). J Gen Comp Endocrinol 26:468–477

Balment RJ, Brimble MJ, Forsling ML (1980) Release of oxytocin induced by salt loading and its influence on renal excretion in the male rat. J Physiol (Lond) 308:439–499

Balment RJ, Brimble MJ, Forsling ML (1982) Oxytocin release and renal actions in normal and Brattleboro rats. Ann NY Acad Sci 394:241–252

Barber RS, Braude R, Mitchell KG (1955) Studies on milk production of Large White pigs. J Agric Sci Camb 46:97–118

Barker JL, Crayton JW, Nicoll RA (1971) Noradrenaline and acetyl choline responses of supraoptic neurosecretory cells. J Physiol (Lond) 218:19–32

Barnafi L, Croxatto H (1966) Vasopressor and oxytocic activities in the posterior lobe of the pituitary gland of male and female rats following adrenalectomy and gonadectomy. Acta Endocrinol 52:3–6

Bell GH, Robson JM (1937) The oxytocin content of the foetal pituitary. QU J Exp Physiol 2:205–208

Benirschke K, Mackay DG (1953) The antidiuretic hormone in fetus and infant. Histochemical observations with special reference to amniotic fluid formation. Obstet Gynecol 1:639–649

Bereznev AP (1963) Oxytocin is a stimulator in the semen production of bulls (in Russian). Vesta Sel Hez Neuk (Mosk) 8:91–95. English translation: Anim Breed Abst (1965) 33:67

Bicknell RJ, Leng G (1982) Endogenous opiates regulate oxytocin but not vasopressin secretion from the neurohypophysis. Nature 298:161–162

Bicknell RJ, Ingram CD, Leng G (1983) Oxytocin release is inhibited by opiates from the neural lobe not those from the intermediate lobe. Neurosci Lett 43:227–230

Bisset GW, Clark BJ, Haldar J (1970) Blood levels of oxytocin and vasopressin during suckling in the rabbit and the problem of their independent release. J Physiol (Lond) 206:711–722
Boer GJ, Boer K, Swaab DF (1982) On the reproductive and developmental differences within the Brattleboro strain. Ann NY Acad Sci 394-45
Bridges TE, Hillhouse EW, Jones MT (1976) The effect of dopamine on neurohypophysial hormone release in vivo and from the rat neural lobe and hypothalamus in vitro. J Physiol (Lond) 260:647–666
Brimble MJ, Dyball REJ (1977) Characterisation of the responses of oxytocin and vasopressin-secreting neurones in the supraoptic nucleus to osmotic stimulation. J Physiol (Lond) 271:253–271
Brimble MJ, Dyball REJ, Forsling ML (1978) Oxytocin release following osmotic activation of oxytocin neurones in the paraventricular and supraoptic nuclei. J Physiol (Lond) 278:69–87
Burford GD, Robinson IC (1982) Oxytocin, vasopressin and neurophysin in the hypothalamo-hypophysial system of the human fetus. J Endocrinol 95:403–408
Burns PJ, Kumaresan P, Douglas RH (1981) Plasma oxytocin concentrations in cyclic mares and sexually aroused stallions. Theriogenology 16:531–539
Burton AM, Forsling ML (1972) Hormone content of the neurohypophysis in foetal, newborn and adult guinea pigs. J Physiol (Lond) 221:6–7P
Chard T, Forsling ML (1976) Bioassay and radioimmunoassay of oxytocin and vasopressin. In: Antoniades HN (ed) Hormones in human plasma. Harvard University Press, Cambridge, pp 489–516
Chard T, Gibbens GLD (1983) Spurt release of oxytocin during surgical induction of labour in women. Am J Obstet Gynecol 147:678–680
Chard T, Boyd NRH, Forsling ML, McNeilly AS, Landon J (1970) The development of a radioimmunoassay for oxytocin. The extraction of oxytocin from plasma and its measurement during parturition in human and goat blood. J Endocrinol 48:223–234
Chard T, Hudson CN, Edwards CRW, Boyd NRH (1971) Release of oxytocin and vasopressin by the human fetus during labour. Nature 234:352–354
Chiesa F, Colombo G, Oberosher R, Seren E (1967) Motility of the bovine uterus during milking. Atti Soc Ital Sci Vet 21:266–269
Clarke G, Lincoln DW (1975) Evidence for a dopaminergic component in the milk ejection reflex of the rat. J Endocrinol 76:32–33P
Clarke G, Fall CHD, Lincoln DW, Merrick LP (1978) Effects of cholinoceptor antagonists on the suckling induced and experimentally evoked release of oxytocin. Br J Pharmacol 63:519–527
Clarke G, Lincoln DW, Merrick LP (1979a) Dopaminergic control of oxytocin release in lactating rats. J Endocrinol 83:409–420
Clarke G, Wood P, Merrick L, Lincoln DW (1979b) Opiate inhibition of peptide release from neurohumoral terminals of hypothalamic neurones. Nature 282:746–748
Cleverley JD (1968) The detection of oxytocin release in response to conditioned stimuli associated with machine milking in the cow. J Endocrinol 40:2–3
Cobo E, De Bernal MM, Gaitan E, Quinetro CA (1967) Neurohypophysial hormone release in the human. II. Experimental study during lactation. Am J Obstet Gynecol 97:519–529
Craft IL, Scrivener R, Dewhurst CJ (1973) Prostaglandin $F_{2\alpha}$ levels in the maternal and fetal circulations in late pregnancy. J Obstet Gynaecol Br Commonw 80:616–618
Cross BA (1955) Neurohormonal mechanisms in emotional inhibition of milk ejection. J Endocrinol 12:29–37
Cross BA (1958) On the mechanism of labour in the rabbit. J Endocrinol 16:261–267
Cross BA, Dyball REJ (1974) Central pathways for neurohypophysial hormone release. In: Geiger SR (ed) Handbook of physiology, sect 7, vol 4, part 1. American Physiological Society, Washington, pp 269–285
Cross BA, Glover TD (1958) The hypothalamus and seminal emission. J Endocrinol 16:385–395

Crowley WR, O'Donohue TL, George JM, Jacobowitz DM (1978) Changes in pituitary oxytocin and vasopressin during the estrus cycle and after ovarian hormones; evidence for mediation by norepinephrine. Life Sci 23:2579–2586
Dale HH (1906) On some physiological actions of ergot. J Physiol (Lond) 34:163–206
Dawood MY, Raghaven KS, Pociask C (1978a) Radioimmunoassay of oxytocin. J Endocrinol 76:261–270
Dawood MY, Raghaven KS, Pociask C, Fuchs F (1978b) Oxytocin in human pregnancy and parturition. Obstet Gynecol 51:138–143
Dawood MY, Wang C, Gupta R, Fuchs F (1978c) Fetal contribution to oxytocin in human labour. Obstet Gynecol 52:205–209
Dawood MY, Ylikorkala O, Trivedi D, Fuchs F (1979) Oxytocin in maternal circulation and amniotic fluid during pregnancy. J Clin Endocrinol Metab 49:429–434
Dawood MY, Khan-Dawood FS, Wahl RS, Fuchs F (1981) Oxytocin release and plasma anterior pituitary and gonadal hormones in women during lactation. J Clin Endocrinol Metab 50:678–683
Debackere M, Peeters G (1960) The influence of vaginal distension on milk ejection and diuresis in the lactating cow. Arch Int Pharmacodyn Ther 123:462–471
Debackere M, Peeters G, Tuyttens N (1961) Reflex release of an oxytocic hormone by stimulation of genital organs in male and female sheep studied by cross circulation technique. J Endocrinol 22:206–214
Dicker SE, Tyler C (1953) Vasopressin and oxytocin activities of the pituitary glands of rats, guinea pigs and cats and of human fetuses. J Physiol (Lond) 121:206–214
Dogterom J, Van Wimersma Greidanus TB, Swaab DF (1977) Evidence for the release of vasopressin and oxytocin into cerebrospinal fluid; measurements in plasma and CSF of intact and hypophysectomised rats. Neuroendocrinology 24:108–118
Donker JD (1958) Lactation studies. I Effects upon milk ejection in the bovine of various injection treatments using oxytocin and relaxin. J Dairy Sci 41:537–544
Dreifuss JJ, Kelly JS (1972) The activity of identified supraoptic neurones and their response to acetyl choline applied by iontophoresis. J Physiol (Lond) 220:105–118
Driesche RV (1961) The nature of the oxytocic activity of human serum during pregnancy. In: Caldeyro-Barcia R, Heller H (eds) Oxytocin. Pergamon, New York, p 400
Edwards BR, La Rochelle FT, Gellai M (1982) Concentration of urine by dehydrated Brattleboro homozygotes: is there a role for oxytocin. Ann NY Acad Sci 394:497–502
Ellendorff F, Forsling ML, Parvizi N, Williams H, Taverne MAM, Smidt D (1979) Plasma oxytocin and vasopressin concentrations in response to prostaglandin injections in the pig. J Reprod Fertil 56:573–577
Ellendorff F, Forsling ML, Poulain DA (1982) The milk ejection reflex in the pig. J Physiol (Lond) 33:577–594
Ely F, Petersen WE (1941) Factors involved in the ejection of milk. J Dairy Sci 24:211–223
Fabian MI, Forsling ML, Jones JJ, Lee J (1969) The release, clearance and plasma protein binding of oxytocin in the anaesthetised rat. J Endocrinol 43:175–189
Falconer J, Mitchell MD, Mountford LA, Robinson JS (1980) Plasma oxytocin concentrations during the menstrual cycle in the rhesus monkey Macaca mulatta. J Reprod Fertil 59:69–72
Fendler K (1961) The effect of endocrine factors on the oxytocic activity of the pituitary. Acta Physiol Acad Sci Hung 20:89–92
Findlay ALR (1966) Sensory discharges in lactating mammary glands. Nature 211:1183–1184
Findlay ALR (1968) The effect of teat anaesthesia on the milk ejection reflex in the rabbit. J Endocrinol 40:127–128
Fitzpatrick RJ (1961) Estimation of small amounts of oxytocin in blood. In: Caldeyro-Barcia R, Heller H (eds) Oxytocin. Pergamon, Oxford, pp 360–379
Fitzpatrick RJ (1966) The posterior pituitary gland and the female reproductive tract. In: Harris GW, Donovan BT (eds) The pituitary gland. Butterworths, London, pp 453–504

Fitzpatrick RJ, Walmsley CF (1965) The release of oxytocin during parturition. In: Pinkerton J (ed) Advances in oxytocin. Pergamon, Oxford
Fliers E, Swaab DF (1983) Activation of vasopressinergic and oxytocinergic neurons during aging in the Wistar rat. Peptides 4:165–170
Flint AP, Forsling ML, Mitchell MD, Turnbull AC (1975) Temporal relationships between changes in oxytocin and prostaglandin in the puerperal ewe. J Reprod Fertil 43:550–554
Flint AP, Forsling ML, Mitchell MD (1978) Blockade of the Ferguson reflex by lumbar epidural anaesthesia in the parturient sheep: effects on oxytocin and uterine venous prostaglandin F levels. Horm Metab Res 10:545–547
Folley SJ (1969) The milk ejection reflex; a neuroendocrine theme in biology, myth and art. J Endocrinol 44:x–xx
Folley SJ, Knaggs GS (1965) Levels of oxytocin in the jugular vein blood of goats during parturition. J Endocrinol 33:301–315
Folley SJ, Knaggs GS (1966) Milk ejection activity (oxytocin) in the external jugular vein blood of the cow, goat and sow in relation to the stimulus of milking and suckling. J Endocrinol 34:197–214
Forsling ML (1970) A comparison of biological and radioimmunological assays of neurohypophysial hormones. Abstracts of Regional Congress of IUPS, pp 316–317
Forsling ML (1976) Antidiuretic hormone, vol 1. Eden, Montreal
Forsling ML (1977) Antidiuretic hormone, vol 2. Churchill Livingstone, Edinburgh, pp 121–128
Forsling ML (1982) Antidiuretic hormone, vol 5. Eden, Montreal
Forsling ML, Brimble MJ (1985) The role of oxytocin in salt and water balance. In: Amico JA, Robinson AG (eds) Oxytocin. Elsevier, Amsterdam, pp 167–175
Forsling ML, Chard T (1979) Exogenous and endogenous posterior pituitary hormones and their effects on the fetus and newborn. In: Laron Z (ed) The influence of maternal hormones on the fetus and newborn, vol 5. Pediatric and adolescent endocrinology. Karger, Basel, pp 36–45
Forsling ML, Martin MJ, Sturdy JC, Burton AM (1973) Observations on the release and clearance of neurophysin and neurohypophysial hormones in the rat. J Endocrinol 57:307–315
Forsling ML, Jack PMB, Nathanielsz PW (1975) Plasma oxytocin concentrations in the fetal sheep. Horm Metab Res 7:198
Forsling ML, Macdonald AA, Ellendorff F (1979 a) The neurohypophysial hormones. Anim Reprod Sci 2:43–56
Forsling ML, Taverne MAM, Parvizi N, Elsaesser F, Smidt D, Ellendorff F (1979 b) Plasma oxytocin concentrations during late pregnancy and lactation in the miniature pig. J Endocrinol 82:61–69
Fox CA, Knaggs GS (1969) Milk-ejection activity (oxytocin) in peripheral venous blood in man during lactation and in association with coitus. J Endocrinol 45:145–146
Freund-Mercier MJ, Richard P (1984) Electrophysiological evidence for facilitatory control of oxytocin neurones by oxytocin during suckling in the rat. J Physiol (Lond) 352:447–466
Fuchs AR (1964) Oxytocin and the onset of labour in rabbits. J Endocrinol 30:217
Fuchs AR (1973) Parturition in rabbits and rats. In: Klopper A, Gardner J (eds) Endocrine factors in labour. Cambridge University Press, Cambridge, pp 163–185
Fuchs AR, Dawood MY (1980) Oxytocin release and uterine activation during parturition in rabbits. Endocrinology 107:1117–1126
Fuchs AR, Cubile L, Dawood MY (1981) Effects of mating on levels of oxytocin and prolactin in the plasma of male and female rabbits. J Endocrinol 90:245–253
Fuchs AR, Husslein P, Sumulong L, Micha JP, Dawood MY, Fuchs F (1982) Plasma levels of oxytocin, 13,14dihydro-15-ketoprostaglandin $F_{2\alpha}$ in preterm labour and the effect of ethanol and ritodrine. Am J Obstet Gynecol 144:753–759
Fuchs AR, Cubile L, Dawood MY, Jorgensen FS (1984) Release of oxytocin and prolactin by rabbits throughout lactation. Endocrinology 114:462–469

Garcia-Villar R, Schams D, Alvinerie M, Laurentie MP, Toutain PL (1985) Activity of the genital tract and plasma levels of oxytocin and cortisol at the time of mating in the ewe. J Endocrinol 105:323–329
Gazarek F, Pohanka J, Lalas M, Fingerova H, Janouskova M, Krikal Z, Hamal Z (1976) Plasma oxytocin and oxytocinase levels in the third trimester of pregnancy and at labour. Endocrinol Exp (Bratisl) 10:283–287
George JM (1976) Vasopressin and oxytocin are depleted from rat hypothalamic nuclei after hypertonic saline. Science 193:146–148
Gibbens GLD, Chard I (1976) Observations on maternal oxytocin release during human labour and the effect of intravenous alcohol administration. Am J Obstet Gynecol 126:243–248
Gibbs DM (1984) Dissociation of oxytocin, vasopressin and corticotrophin secretion during different types of stress. Life Sci 35:487–492
Gill JC, Thompson W (1965) Observations on the behaviour of suckling pigs. Br J Anim Behav 4:46–51
Gillespie A, Brummer HC, Chard T (1972) Oxytocin release by infused prostaglandins. Br Med J 1:543–544
Ginsburg M (1968) Production, release transportation and elimination of the neurohypophysial hormones. In: Berde B (ed) Neurohypophysial hormones and similar peptides. Springer, Berlin Heidelberg New York, pp 286–371 (Handbook of experimental pharmacology, vol 23)
Glatz TH, Weitzman RE, Eliot RH, Klein AH, Nathanielsz PW, Fisher DA (1981) Ovine maternal and fetal plasma oxytocin concentrations before and during parturition. Endocrinology 108:1328–1332
Goodfellow CF, Hull MGR, Swaab DF, Dogterom J, Buijs RM (1985) Oxytocin deficiency at delivery with epidural anaesthesia. Br J Obstet Gynaecol 90:214–219
Gorewit RC, Wachs EA, Sagi R, Merill WG (1983) Current concepts on the role of oxytocin in milk ejection. J Dairy Sci 66:2236–2250
Gunther M (1948) The posterior pituitary and labour. Br Med J 1:567
Guzek JW, Janus J, Gosek J (1981) The vasopressor and oxytocic activities of the hypothalamus and neurohypophysis are influenced by inhibited alpha-adrenergic transmission during dehydration and subsequent rehydration in the white rat. Acta Physiol Pol 32:667–679
Haldar J (1970) Independent release of oxytocin and vasopressin during parturition in the rabbit. J Physiol (Lond) 206:723–730
Haldar J, Sawyer WH (1978) Inhibition of oxytocin release by morphine and its analogs. Proc Soc Exp Biol Med 157:476–480
Harris GW (1947) The innervation and actions of the neurohypophysis: an investigation using the method of remote control stimulation. Philos Trans R Soc Lond [Biol] 232:385–344
Harris GW, Pickles VP (1953) Reflex stimulation of the neurohypophysis (posterior pituitary gland) and the nature of the posterior pituitary hormone(s). Nature 172:1049
Hawker RW (1961) Oxytocin and an unidentified oxytocic substance in extracts of blood. In: Caleyro-Barcia R, Heller H (eds) Oxytocin. Pergamon, Oxford, pp 425–434
Hawker RW, Robertson PH (1958) Some properties of an oxytocic substance found in blood extracts. Endocrinology 63:242–249
Heller H (1958) The neurohypophysis during the oestrus cycle, pregnancy and lactation. In: Lloyd CW (ed) Recent progress in the endocrinology of reproduction. Academic, New York, pp 365–387
Higuchi T, Honda K, Fukuoka T, Negoro H, Wakabayashi K (1985) Release of oxytocin during suckling and parturition in the rat. J Endocrinol 105:339–346
Homeida AM, Cooke RG (1983) Suppression of pulsatile release of oxytocin during early pregnancy in the goat. Prostaglandins 26:103–109
Houvenaghel A, Peeters G, Djordjevic N (1967) Afwezigheid van ocytocineintscheiding tijdens de coitus bij ram en bok. Vlaams Diergenesskundig Tijdschrift 36:605–611
Huntingford PJ (1961) Intranasal use of synthetic oxytocin in management of breast feeding. Br Med J 1:709–711

Jackson R, George JM (1980) Oxytocin in microdissected hypothalamic nuclei; significant differences between prepubertal and sexually mature female rats. Neuroendocrinology 31:158–160
Jakubowksa-Naziemblo B (1979) Oxytocic activity of plasma and posterior lobe after ovariectomy and implantation of stilboestrol or progesterone tablets in female rats. Acta Physiol Pol 30:597–606
Jones CW, Pickering BT (1969) Comparison of the effects of water deprivation and sodium chloride inhibition on the neurohypophysis of the rat. J Physiol (Lond) 203:469–458
Juss TS, Wakerley JB (1981) Mesencephalic areas controlling pulsatile oxytocin release in the suckled rat. J Endocrinol 91:233–244
Kalin NH, Gibbs DM, Barksdale CM, Shelton SE, Carnes M (1985) Behavioral stress decrease plasma oxytocin concentrations in primates. Life Sci 36:1275–1280
Khan-Dawood FS, Dawood MY (1983) Human ovaries contain immunoreactive oxytocin. J Clin Endocrinol Metab 57:1129–1132
Khan-Dawood FS, Dawood MY (1984) Oxytocin content of human fetal pituitary glands. Am J Obstet Gynecol 148:420–423
Knaggs GS (1963) Blood oxytocin levels in the cow during milking and in the parturient goat. J Endocrinol 26:xxiv–xxv
Konig A, Bottcher D (1966) Der Hormongehalt des Hypophysen-Hinterlappens während des Sexualzyklus von Wistar Ratten. Arch Gynekol 203:485–490
Konig A, Ehlers B (1973) Effect of stilboestrol on hormonal activity of the posterior lobe of mature female rats. Rass Neurol Veg 22:193–201
Krnjevic K, Phillis JW (1963) Pharmacological properties of acetyl choline-sensitive cells in the cerebral cortex. J Physiol (Lond) 166:328–350
Kumar MSA, Chen CL, Muther TF (1979) Changes in the pituitary and hypothalamic content of methionine enkephalin during the oestrus cycle of rats. Life Sci 25:1687–1696
Kumaresan P, Anandarangam PB, Dianzon W, Vasicka A (1974) Plasma oxytocin levels during human pregnancy and labour as determined by radioimmunoassay. Am J Obstet Gynecol 119:215–233
Kumaresan P, Han GS, Anandarangam PB, Vasicka A (1975) Oxytocin in maternal and fetal blood. Obstet Gynecol 46:272–274
Kumaresan P, Weathersbee PS, Lodge JR (1979) Effect of ejaculation on plasma oxytocin levels in dairy bulls. Fed Proc 38:979
Kumaresan P, Kumaresan M, Hossini M, Avelland C, Vasicka A (1983) Human ovulation and plasma oxytocin. Int J Gynaecol Obstet 21:413–418
Landgraf R, Gunther O (1983) Vasopressin and oxytocin in cerebrospinal fluid and plasma of conscious rabbits – response to dehydration and haemorrhage. Biomed Biochim Acta 42:1339–1341
Landgraf R, Hacker R, Buhl H (1982a) Plasma vasopressin and oxytocin response to exercise and during a day-night cycle in man. Endokrinologie 79:281–291
Landgraf R, Wehowsky G, Schulz J, Schulze H, Bothur D (1982b) Radioimmunoassay measurement of plasma oxytocin and vasopressin in cows during machine milking. Endokrinologie 79:434–436
Lang RE, Heil J, Ganten D, Hermann K, Unger T, Rascher W (1983) Oxytocin unlike vasopressin is a stress hormone in the rat. Neuroendocrinology 39:314–316
Lawson DM, Graf GC (1968) Plasma oxytocic activity and intramammary pressure in lactating dairy cows. J Dairy Sci 51:1676–1679
Leake RD, Weitzman RE, Fisher DA (1981a) Oxytocin concentrations during the neonatal period. Biol Neonate 39:127–131
Leake RD, Weitzman RE, Glatz TH, Fisher DA (1981b) Plasma oxytocin concentrations in men, non-pregnant women and pregnant women before and during spontaneous labour. J Clin Endocrinol Metab 53:730–733
Leake RD, Waters CB, Rubin T, Buster JE, Fisher DA (1983) Oxytocin and prolactin responses in long term breast feeding. Obstet Gynecol 62:565–568
Leake RD, Buster JE, Fisher DA (1984) The oxytocin secretory response to breast stimulation in women during the menstrual cycle. Am J Obstet Gynecol 148:457–460

Lefcourt AM, Akers RM (1984) Small increases in peripheral noradrenaline inhibit the milk ejection response by means of a peripheral mechanism. J Endocrinol 100:337–344

Legros JJ, Grau JD (1973) Effects of ethinyl-oestradiol on some neurohypophysial active components in the rat. Nature 241:247–249

Legros JJ, Franchimont P, Burger H (1975) Variations of neurohypophysial function in normally cycling women. J Clin Endocrinol Metab 91:54–59

Lincoln DW (1974) Does a mechanism of negative feedback determine the intermittent release of oxytocin during suckling? J Endocrinol 60:193–194

Lincoln DW, Paisley AC (1982) Neuroendocrine control of milk ejection. Reprod Fertil 65:571–586

Lincoln DW, Wakerley JB (1974) Electrophysiological evidence for the activation of supraoptic neurosecretory cells during release of oxytocin. J Physiol (Lond) 242:533–584

Lincoln DW, Wakerley JB (1975) Factors governing the periodic activation of supraoptic and paraventricular neurosecretory cells during suckling in the rat. J Physiol (Lond) 250:443–461

Lincoln DW, Hill A, Wakerley JB (1973) The milk ejection reflex of the rat: an intermittent function not abolished by surgical levels of anaesthesia. J Endocrinol 57:459–478

Lincoln DW, Hentzenk, Hin T, Van der Schoot P, Clark G, Summerlee AJS (1980) Sleep a prerequisite for reflex milk ejection in the rat. Exp Brain Res 38:151–162

Lucas A, Drewett RB, Mitchell MD (1980) Breast feeding and plasma oxytocin concentrations. Br Med J 281:834–835

Macdonald AA, Forsling ML, Williams H, Ellendorff F (1979) Plasma vasopressin and oxytocin concentrations in the conscious pig fetus; response to haemorrhage. J Endocrinol 81:124P–125P

Martin R, Voigt KH (1981) Enkephalins co-exist with oxytocin and vasopressin in nerve terminals of rat neurohypophysis. Nature 389:502–504

Mason WT (1983) Excitation by dopamine of putative oxytocinergic neurones in rat supraoptic nucleus in vitro: evidence for two classes of continuously firing neurones. Brain Res 267:113–121

McNeilly AS (1972) The blood levels of oxytocin during suckling and handmilking in the goat with some observations on the pattern of hormone release. J Endocrinol 52:177–188

McNeilly AS, Ducker HA (1972) Blood levels of oxytocin in the female goat during coitus and its response to stimuli associated with mating. J Endocrinol 54:399–406

McNeilly AS, Folley SJ (1970) Blood levels of milk ejection activity (oxytocin) in the female goat during mating. J Endocrinol 48:9–10

McNeilly AS, McNeilly JR (1978) Spontaneous milk ejection during lactation and its possible relevance to success of breast feeding. Br Med J 2:466–468

McNeilly AS, Robinson ICAF, Houston MJ, Howie PW (1983) Release of oxytocin and prolactin in response to suckling. Br Med J 286:257–259

Mitchell MD, Webb R, Brennecke SP, Anderson ABM, Turnbull AC (1972) Oxytocin in the fetal and maternal circulations during induced and spontaneous parturition in sheep. Eur J Obstet Gynecol Reprod Biol 14:115–120

Mitchell MD, Haynes RJ, Anderson ABM, Turnbull AC (1980a) Oxytocin in human ovulation. Lancet 2:704

Mitchell MD, Mountford LA, Natale R, Robinson JS (1980b) Concentrations of oxytocin in the plasma and amniotic fluid of the rhesus monkey (Macaca mulatta) during the latter half of pregnancy. J Endocrinol 84:473–478

Miyake A, Tasaka K, Sakumoto T, Aono T (1983) Arginine vasopressin suppresses oxytocin release from the superfused hypothalamo-pituitary complex. Acta Endocrinol 103:497–500

Mizuno H, Talwalker PK, Meites J (1967) Central inhibition by serotonin of reflex release of oxytocin in response to suckling stimulus in the rat. Neuroendocrinology 2:222–231

Momongan VG, Schmidt GH (1970) Oxytocin levels in the plasma of Holstein Friesian cows during milking with and without a premilking stimulus. J Dairy Sci 53:747–751

Moos F, Richard P (1975) Role de la noradrenaline et de l'acetyl choline dans la liberation d'ocytocine induite par des stimulations vaginale vagale et mammaire. J Physiol (Paris) 70:315–322
Moos F, Richard P (1979a) The inhibitory role of β-noradrenergic receptors in oxytocin release during suckling. Brain Res 169:595–599
Moos F, Richard P (1979b) Effects of dopaminergic agonist and antagonist on oxytocin release induced by various stimuli. Neuroendocrinology 28:138–144
Moos F, Richard P (1982) Excitatory effects of dopamine on oxytocin and vasopressin reflex releases in the rat. Brain Res 241:249–260
Moos F, Richard P (1983) Serotonergic control of oxytocin release during suckling in the rat: opposite effects in conscious and anaesthetised rats. Neuroendocrinology 36:300–306
Morris JF (1982) The Brattleboro magnocellular neurosecretory system: a model for the study of peptidergic neurons. Ann NY Acad Sci 394:54–69
Mosdel G, Sjaastad OV, Bloom AK (1981) Plasma concentrations of oxytocin and intramammary pressure in goats during manual stimulation of the udder and hand milking. J Endocrinol 90:159–166
Moss RL, Urban I, Cross BA (1972) Microelectrophoresis of cholinergic and aminergic drugs on paraventricular neurones. Am J Physiol 223:310–318
Negoro H, Visessuwan S, Holland RC (1973) Reflex activation of paraventricular nucleus units during the reproductive cycle in ovariectomised rats treated with oestrogen or progesterone. J Endocrinol 59:559–567
Newton M, Egli GE (1958) The effect of intranasal administration of oxytocin on the let down of milk in lactating women. Am J Obstet Gynecol 76:103–107
Nicholson HD, Swann RW, Burford GD, Wathes DC, Porter DG, Pickering BT (1983) More steroidogenic tissues containing oxytocin and vasopressin: human testis and fetal adrenal. Acta Endocrinol 103 [Suppl 256]:243
Nordman JJ, Dreifuss JJ (1972) Hormone release evoked by electrical stimulation of rat neurohypophysis in the absence of action potentials. Brain Res 45:604–607
Otsuki Y, Tanizawa O, Yamiji K, Fujita M, Kurachi K (1983) Feto-maternal plasma oxytocin levels in normal and anenkephaliç pregnancies. Acta Obstet Gynecol Scand 62:235–237
Ott I, Scott JC (1910) The action of infundibulin upon the mammary secretion. Proc Soc Exp Biol Med 8:48–49
Parkash V, Anderson RR (1972) Pituitary and blood plasma oxytocic activity and oxytocin disappearance rates in cattle. J Dairy Sci 53:75–79
Pearson DB, Goodman R, Sachs H (1975) Stimulated vasopressin synthesis by a fetal hypothalamic factor. Science 187:1081–1082
Peeters G, Stormorken H, Vanschoubroek F (1960) The effect of different stimuli on milk-ejection and diuresis in the lactating cow. J Endocrinol 20:163–172
Peeters G, Legros JJ, Piron Bossuyt C, Reynaert R, Vanden Driessche R, Vannieuwenhuyses (1983) Release of neurophysin I and oxytocin by stimulation of the genital organs in bulls. J Endocrinol 99:161–171
Perlow MJ, Reppert SM, Artman HA, Fisher DA, Seif SM, Robinson AG (1982) Oxytocin vasopressin and oestrogen stimulated neurophysin: daily patterns of secretion in cerebrospinal fluid. Science 216:1416–1418
Piecuwiak H, Jantzen V, Konig A (1976) The influence of progesterone on the antidiuretic and oxytocic activities of the posterior lobe of female rats. Arch Gynekol 221:139–144
Pitzel L, Bischoff C, Konig A (1981) Changes in posterior pituitary oxytocin release in vitro during the oestrus cycle of female rats. Exp Brain Res 44:229–231
Poulain DA, Wakerley JB, Dyball REJ (1977) Electrophysiological differentiation of oxytocin- and vasopressin-secreting neurons. Proc R Soc Lond [Biol] 196:367–384
Poulain DA, Rodriguez A, Ellendorff F (1981) Sleep is not a prerequisite for the milk ejection in the pig. Exp Brain Res 43:107–110
Prilusky J, Deis RP (1975) Effect of L-DOPA on milk ejection and prolactin release in lactating rats. J Endocrinol 67:397–401

Pritchard DE, Hays RL (1966) Oxytocin concentration in bovine blood plasma during normal milkings. J Dairy Sci 47:736–737
Randy HA, Graf GG, Bibb TL (1971) Relationship between plasma oxytocic activity and reproductive status of the lactating bovine. J Dairy Sci 54:779–784
Reid RL, Yen SSC, Artman H, Fisher DA (1981) Effects of synthetic β-endorphin on release of neurohypophyseal hormones. Lancet 2:1169–1170
Reppert SM, Schwartz WJ, Artman HG, Fisher DA (1983) Comparison of temporal profiles of vasopressin and oxytocin in the CSF of cat, monkey and rat. Brain Res 261:341–345
Richard P (1970) An electrophysiological study in the ewe of the tracts which transmit impulses from the mammary glands to the pituitary stalk. J Endocrinol 47:37–44
Roberts JS (1973) Cyclical fluctuations in reflexive oxytocin release during the estrus cycle of the goat. Biol Reprod 13:314–317
Roberts JS, Share L (1968) Effects of progesterone and oestrogen on blood levels of oxytocin during vaginal distension. Endocrinology 83:272–278
Roberts JS, Share L (1970) Inhibition by progesterone of oxytocin release during vaginal distension. Endocrinology 87:812–815
Robinson AG, Ferin M, Zimmerman EA (1976) Plasma neurophysin levels in monkeys emphasis on the hypothalamic response to oestrogen and ovarian events. Endocrinology 98:468–475
Robinson AG, Haluszczak C, Wilkins J, Huellmantel AB, Watson CG (1977) Independent physiologic control of two neurophysins in humans. J Clin Endocrinol Metab 44:330–339
Robinson ICAF, Jones PM (1982) Oxytocin and neurophysin in plasma and CSF during suckling in the guinea pig. Neuroendocrinology 34:59–63
Robinson ICAF, Clarke RG, Fairhall KM, Jones PM, Parsons JA (1982) Effects of oxytocin antiserum in Brattleboro rats. Ann NY Acad Sci 394:285–290
Robinson JE, Short RV (1977) Changes in breast sensitivity at puberty, during the menstrual cycle and at parturition. Br Med J 1:1188–1191
Rossier J, Battenberg E, Pittman Q, Bayon A, Koda L, Miller R, Guillemin R, Bloom F (1979) Hypothalamic enkephalin neurones may regulate the neurohypophysis. Nature 277:653–655
Ryden G, Sjoholm I (1969) Half life of oxytocin in blood of pregnant and non-pregnant woman. Acta Endocrinol (Copenh) 61:425–431
Sala NL, Luther EC, Arballo JC, Cordero-Funes JC (1974) Oxytocin reproducing reflex milk ejection in lactating women. J Appl Physiol 36:154–158
Schams D (1983) Oxytocin determination by radioimmunoassay. III Improvement to subpicogram sensitivity and application to blood levels in cyclic cattle. Acta Endocrinol 103:180–183
Schams D, Schmidt Polex B, Kruse V (1979) Oxytocin determination by radioimmunoassay in cattle 1. method and preliminary physiological data. Acta Endocrinol 92:258–270
Schams D, Baumann G, Leidl W (1982a) Oxytocin determination by radioimmunoassay in cattle. II Effect of mating and stimulation of genital tract in bulls, cows and heifers. Acta Endocrinol 99:218–223
Schams D, Lahou-Kassi A, Glatzel P (1982b) Oxytocin concentrations in peripheral blood during the oestrus cycle and after ovariectomy in two breeds of sheep with low and high fecundity. J Endocrinol 92:9–13
Schams D, Mayer H, Prokopp A, Worstoff H (1984) Ovarian secretion of oxytocin during milking in dairy cows with regard to the variation and importance of a threshold level for milk removal. J Endocrinol 102:337–343
Schriefer JA, Miller JW (1979) Investigation of the role of fetal oxytocin during parturition in the rat. Fed Proc 38:270
Schriefer JA, Lewis PR, Miller JW (1980) Effect of dopamine on length of gestation and on the release of fetal oxytocin in rats. J Pharmacol Exp Ther 212:431–434
Sellars SM, Hodgson HT, Mountford LA, Mitchell MD, Anderson ABM, Turnbull AC (1981) Is oxytocin involved in parturition. Br J Obstet Gynaecol 88:725–729

Seybold V, Miller TW, Lewis PR (1976) Investigation of a dopaminergic mechanism for regulating oxytocin release. J Pharmacol Exp Ther 207:605–610
Sharma SC, Chaudhury RR (1970) Studies on mating, part 1. The absence of release of oxytocin at mating in female rabbits. Indian J Med Res 58:495–500
Sharma SC, Hays RL (1968) Release of oxytocin due to genital stimulation in bulls. J Dairy Sci 7:966–969
Sharma SC, Fitzpatrick RJ, Ward WR (1972) Coital-induced release of oxytocin in the ram. J Reprod Fertil 31:488–489
Sheldrick EL, Flint APF (1981) Circulating concentrations of oxytocin during the estrous cycle and early pregnancy in the sheep. Prostaglandins 22:631–636
Silverman AJ (1975) The hypothalamic magnocellular neurosecretory system of the guinea pig. II Immunohistochemical localisation of neurophysin and vasopressin in the fetus. Am J Anat 144:445–460
Sinding C, Seif SM, Robinson AG (1980) Levels of neurohypophyseal peptides in the rat during the first month of life. I Basal levels in plasma, pituitary and hypothalamus. Endocrinology 107:749–754
Skowsky WR, Fisher DA (1977) Fetal neurohypophysial arginine vasopressin and arginine vasotocin in man and sheep. Pediatr Res 11:627–630
Skowsky WR, Swan L, Smith P (1979) Effects of sex steroids on arginine vasopressin in intact and castrated male and female rats. Endocrinology 104:105–108
Stegner H, Leake RD, Palmer SM, Oakes G, Fisher DA (1984) The effect of hypoxia on neurohypophyseal hormone release in fetal and maternal sheep. Pediatr Res 18:188–191
Sterba B (1979) Extrahypothalamic vasopressinergic and oxytocinergic pathways. In: Laszlo FA (ed) Recent results in peptide hormone and androgenic steroid research. Akademiai, Budapest, pp 102–110
Summerlee AJS (1983) Hypothalamic neurone activity hormone release and behaviour in freely moving rats. Qu J Exp Physiol 68:505–515
Summerlee AJS, Lincoln DW (1981) Electrophysiological recordings from oxytocinergic neurones during milk ejection in the unanaesthetised lactating rat. J Endocrinol 90:225–265
Summerlee AJS, O Bryne KT, Paisley AC, Breeze MF, Porter DG (1984) Relaxin affects the central control of oxytocin release. Nature 309:372–374
Sundsten JW, Sawyer CH (1961) Osmotic activation of neurohypophysial hormone release in rabbits with hypothalamic islands. Exp Neurol 4:548–561
Swaab DF, Jongkind JF (1970) The hypothalamic neurosecretory activity during the oestrus cycle, pregnancy, parturition, lactation and persistent oestrus and after gonadectomy in the rat. Neuroendocrinology 6:135–345
Swaab DF, Jongkind JF (1971) Influence of gonadotrophic hormones on the hypothalamic neurosecretory activity in the rat. Neuroendocrinology 8:36–47
Sybesma W (1968) Oxytocic activity measurements in blood plasma of cows and bulls. VI Int Congr Anim Reprod 1:825–828, Paris
Taverne MAM, Naaktgeboren C, Elsaesser F, Forsling ML, Van der Weyden GC, Ellendorff F, Smidt D (1979) Myometrial electrical activity and plasma concentrations of progesterone, oestrogens and oxytocin during late pregnancy, parturition and the first week post partum in the miniature pig. Biol Reprod 21:1125–1143
Thomas TH, Lee MR (1976) The specificity of antisera for the radioimmunoassay of arginine vasopressin in human plasma and urine during water loading and dehydration. Clin Sci Mol Med 51:525–536
Tindal JS (1975) Stimuli that cause release of oxytocin. In: Geiger SR (ed) Handbook of physiology, sect 7, vol 4, part 1. American Physiological Society, Washington, pp 257–267
Tindal JS, Knaggs GS (1971) Determination of the detailed hypothalamic route of milk ejection reflex in the guinea pig. J Endocrinol 50:135–152
Tindal JS, Knaggs GS (1975) Further studies on the afferent path of the milk ejection reflex in the brain stem of the rabbit. J Endocrinol 66:107–113
Tindal JS, Knaggs GS, Turvey A (1967) The afferent path of the milk ejection reflex in the brain of the guinea pig. J Endocrinol 38:337–349

Tribollet E, Clarke G, Dreifuss JJ, Lincoln DW (1978) The role of central adrenergic receptors in the reflex release of oxytocin. Brain Res 142:69–84
Urban I, Moss RL, Cross BA (1971) Problems in electrical stimulation of afferent pathways for oxytocin release. J Endocrinol 51:347–358
Valtin H, Schroeder HA (1964) Familial hypothalamic diabetes insipidus in rats (Brattleboro strain). Am J Physiol 206:425–436
Valtin H, Sawyer WH, Sokol HW (1965) Neurohypophysial principles in rats homozygous and heterozygous for hypothalamic diabetes insipidus (Brattleboro strain). Endocrinology 77:701–706
Van Demark NL, Hays RL (1951) Stimulatory action of breeding on the release of oxytocin as measured by intramammary pressure. J Dairy Sci 34:496–497
Van Demark NL, Hays RL (1953) Effect of stimulation of the reproductive organs of the cow on the release of an oxytocin-like substance. Endocrinology 52:634–637
Vanderhaegenhen JJ, Lostra F, Liston DR, Rossier J (1983) Proenkephalin, [Met]enkephalin and oxytocin immunoreactivity are colocalised in bovine hypothalamic magnocellular neurons. Proc Natl Acad Sci (USA) 80:5139–5143
Van Dongen CG (1965) A bioassay for oxytocic activity in biological fluids and its use in certain phases of reproductive biology. Dissert Abst 25:6726
Van Dongan CG, Hays RL (1963) Oxytocic activity in unextracted blood plasma during calving. J Reprod Fertil 11:317–323
Van Wimersma Greidanus TB, De Wied D (1977) The physiology of neurohypophysial system and its relation to memory processes. In: Davidson AN (ed) Biochemical correlates of brain structure and function. Academic, London, pp 215–248
Vasicka A, Kumaresan P, Hans GS, Kumaresan M (1978) Plasma oxytocin in initiation of labour. Am J Obstet Gynecol 130:263–273
Verney EB (1946) Absorption and excretion of water. The antidiuretic hormone. Lancet 2:251, 739–733, 781–783
Verney EB (1947) The antidiuretic hormone and the factors which determine its release. Proc R Soc Lond [Biol] 135:25–106
Vizi ES, Volkebas (1980) Inhibition by dopamine of oxytocin release from isolated posterior lobe of the hypophysis of the rat; disinhibitory effect of β-endorphin/enkephalin. Neuroendocrinology 31:46–52
Vizsolyi E, Perks AM (1976) Neurohypophysial hormones in fetal life and pregnancy. II Chromatographic studies in the sheep (oves aries). Gen Comp Endocrinol 29:41–50
Vorherr H (1979) Effect of prostaglandins on milk ejection. Endokrinologie 74:257–270
Wakerley JB, Lincoln DW (1971) Intermittent release of oxytocin during suckling in the infant rat during spontaneous milk ejection. Physiol Behav 15:277–281
Wakerley JB, Lincoln DW (1973) Milk ejection in the rat: the result of a selective release of oxytocin. J Endocrinol 57:557–558
Walmsley CF (1963) Blood oxytocin levels in relation to reproduction. M Sc Thesis, University of Bristol
Wang KL, Fisher K, Ledens K, North W (1982) The ontogeny of the oxytocin/neurophysin system of the Brattleboro rat. Ann NY Acad Sci 394:177–182
Wathes DC (1984) Possible actions of gonadal oxytocin and vasopressin. J Reprod Fertil 71:315–345
Wathes DC, Swann RW (1982) Is oxytocin an ovarian hormone? Nature 297:225–259
Webb R, Mitchell MD, Falconer J, Robinson JS (1981) Temporal relationship between peripheral plasma concentration of oxytocin, progesterone and 13,14 dihydro-15-keto prostaglandin F_2 during the oestrus cycle and early pregnancy in the ewe. Prostaglandins 22:443–453
Weitzman RE, Glatz TH, Fisher DA (1978) The effect of haemorrhage and hypertonic saline upon plasma oxytocin and vasopressin in conscious dogs. Endocrinology 103:2154–2160
Weitzman RE, Leake RD, Rubin RT, Fisher DA (1980) The effect of nursing on neurohypophysial hormone and prolactin secretion in human subjects. J Clin Endocrinol Metab 51:836–839
Whittlestone WG (1954) Intramammary pressure changes in the lactating sow. I The effect of oxytocin. J Dairy Sci 21:19–30

Proteolytic Conversion of Oxytocin, Vasopressin, and Related Peptides in the Brain

J. P. H. Burbach

Contents

1 Introduction

The nonapeptides oxytocin and vasopressin (arginine-vasopressin or lysine-vasopressin) have dual functions: they are hormones in the periphery and neuropeptides in the brain. As hormones they serve the organism by controlling peripheral functions, e.g., maintenance of water balance and blood pressure, and participate in parturition and lactation. As neuropeptides they play a role in the regulation of memory processes (De Wied 1965, 1971, 1976, 1980, 1983), body temperature (Cooper et al. 1979; Veale et al. 1981; Kasting et al. 1980), blood pressure (Versteeg et al. 1983; Feuerstein et al. 1984), brain development (Boer et al. 1980; Boer 1984), maternal behavior (Pedersen and Prange 1979; Pedersen et al. 1982), sexual behavior (Bohus 1977; Södersten et al. 1983), and development of tolerance and addiction to narcotic drugs (Van Ree and De Wied 1981; Van Ree 1982; Hoffman 1982). In addition, they may have paracrine functions in peripheral tissues such

Rudolf Magnus Institute for Pharmacology, Medical Faculty, Rijksuniversity of Utrecht, Vondellaan 6, 3521 GD Utrecht, The Netherlands

Current Topics in Neuroendocrinology, Vol. 6

OXYTOCIN	H-Cys^1-Tyr-Ile-Gln-Asn-Cys-Pro-Leu-Gly^9-NH_2
Arg-VASOPRESSIN	H-Cys^1-Tyr-Phe-Gln-Asn-Cys-Pro-Arg-Gly^9-NH_2
Lys-VASOPRESSIN	H-Cys^1-Tyr-Phe-Gln-Asn-Cys-Pro-Lys-Gly^9-NH_2
Arg-VASOTOCIN	H-Cys^1-Tyr-Ile-Gln-Asn-Cys-Pro-Arg-Gly^9-NH_2

Fig. 1. Primary structure of oxytocin and related nonapeptides. Structural differences are shown in *boxes*

as the ovary, testis, and adrenal gland (Wathes and Swann 1982; Nicholson et al. 1984; Ang and Jenkins 1984; Lim et al. 1984).

Oxytocin and vasopressin, which are encoded by related but separate genes, are biosynthesized together with their respective neurophysins via processing of a precursor (Land et al. 1982, 1983; Schmale et al. 1983; Ruppert et al. 1984; Schmale and Richter 1984; Ivell and Richter 1984). In the brain the genes of oxytocin and vasopressin are expressed by distinct groups of neurons. These neurons either project to the posterior pituitary and secrete oxytocin and vasopressin into the blood, or innervate a number of brain structures. For instance, vasopressin-containing fibers from the supraoptic nucleus run mainly to the posterior lobe of the pituitary gland, while the suprachiasmatic nucleus innervates only the brain (Buijs 1978; Buijs et al. 1983; Zimmerman 1984; Sofroniew and Weindl 1978; Sofroniew 1980). This organization of peptidergic neurons provides the anatomical basis for the separate peripheral and central roles of oxytocin and vasopressin.

The structural requirements of the peptides for peripheral and central activities are also dissociated. Unlike their peripheral activities, the central activities of oxytocin and vasopressin are not limited to the complete nonapeptide structure per se. For instance, the des-GlyNH_2 derivatives of arginine-vasopressin and lysine-vasopressin are virtually devoid of agonistic vasopressor and antidiuretic activity, but fully retain the central activities associated with vasopressin (De Wied 1976, 1980; De Wied et al. 1972). Moreover, fragments of oxytocin and vasopressin can have central effects of their own, or can be more potent (Burbach et al. 1983 a, b). These findings suggested that oxytocin and vasopressin might be precursor peptides in the brain, and pointed to a functional role of proteolytic events responsible for the generation of peptides with novel biologic properties (De Wied and Bohus 1978; De Wied 1977, 1980; Burbach and De Wied 1981). In addition, other proteolytic processes must provide mechanisms to inactivate oxytocin and vasopressin in order to terminate their action. This chapter focusses on the proteolytic conversion of oxytocin and vasopressin and related peptides (Fig. 1) by brain peptidases.

The proteolysis of oxytocin and related peptides in the brain has been studied since the early 1960s. During the last 2 decades, ideas about the role of oxytocin and vasopressin in the brain, and about the significance of proteolysis for peptide production and degradation, have evolved. While in initial studies proteolysis of oxytocin in the brain was thought to be in some way related to production of the peptide for peripheral functions (e.g., Hooper 1964), others considered inactivation and degradation a major role (e.g., Pliska et al. 1971 a, b). However, oxytocin is one of the first brain peptides for which a function as precursor for active fragments was presumed (e.g., Walter 1972). Now, more detailed information about the routes of conversion of oxytocin and related peptides in the brain and the identity and activities of fragments is available. Data about the proteolytic processing of oxytocin and related peptides, obtained in the last 2 decades, are reviewed here, and the concepts on which the studies were based are discussed.

2 Proteolytic Conversion of Oxytocin

During the last 2 decades, the cleavage of oxytocin by peptidases in the brain has been studied by several groups, who investigated the proteolysis of oxytocin with different aims and therefore different strategies. The pioneering work of Hooper and colleagues in the 1960s was based on the consideration that inactivation of oxytocin in the hypothalamus could have consequences for the amounts of oxytocin available for peripheral functions related to pregnancy and lactation (e.g., Hooper 1964; Frith and Hooper 1974). Consequently, these studies merely aimed to relate hypothalamic oxytocin-inactivating or -degrading enzyme activity to the endocrine status associated with pregnancy and lactation. In the early 1970s, the groups of Celis and Walter considered oxytocin as a precursor peptide for factors influencing the release of α-melanocyte-stimulating hormone (MSH) from the pituitary gland (e.g., Celis and Taleisnik 1971 a, b; Walter 1972). Since the late 1970s, the proteolysis of oxytocin has been studied by Burbach and De Wied, based on the concept that fragments of oxytocin generated by brain peptidases could represent neuropeptides with more selective and/or potent effects on brain functions (Burbach et al. 1980 a, 1983 a, b; Burbach and De Wied 1981). The precursor concept implied to identify the fragments generated by proteolysis and to delineate the routes and mechanisms of their formation.

2.1 Early Observations of Oxytocin Inactivation by Brain Peptidases

The first observations of peptidases in the brain which were able to cleave oxytocin were reported by Hooper (1962). He observed that incubation of oxytocin with homogenates of dog hypothalamus resulted in a loss of oxytocic activity as determined by bioassay. Also, vasopressin, bradykinin, and substance P were inactivated by the homogenate. Peptidase activity was relatively high in the hypothalamus, compared to other brain regions (Hooper 1963). Subcellular fractionation of the hypothalamic homogenate indicated that enzyme activity was present in the soluble fraction, as well as in the mitochondrial and nuclear fractions (Hooper 1964). Further fractionation of the mitochondrial preparation showed enrichment of oxytocin-inactivating activity in the heavy fraction, which could contain lysosomal and synaptic elements in addition to mitochondria. Hooper argued that the particulate and soluble activities were not identical (Hooper 1964).

Celis and Taleisnik (1971 a, b) observed that incubation of median eminence extracts from rats gave rise to two factors that influenced the release of pituitary MSH; one inhibited release, the other stimulated it. Production of these factors was enhanced when either oxytocin or lysine-vasopressin, but not arginine-vasopressin, was coincubated. The biologic activity of oxytocin was reduced. The authors suggested that oxytocin was degraded by peptidase activity, while MSH-release-inhibiting and -stimulating factors were generated (Celis and Taleisnik 1971 a; see also Sect. 2.4).

These studies provided evidence of enzymatic inactivation in the brain, but did not identify cleavage sites or metabolic products.

2.2 Sites of Proteolysis in the Oxytocin Molecule

The above-mentioned studies were limited to observation of the disappearance of oxytocic biologic activity. Loss of activity can be caused by any cleavage in the oxytocin molecule. The synthesis of ^{14}C-labeled oxytocins by Walter and coworkers has been of great importance for the identification of cleavage sites in oxytocin (Walter and Havran 1971; Hoffman and Walter 1976; Stahl and Walter 1978). Using [^{14}C-$GlyNH_2^9$]oxytocin as substrate and high-voltage paper electrophoresis to separate products, it was found after incubation with whole rat brain fractions that ^{14}C-radioactive material comigrated with $GlyNH_2$ and the C-ter-

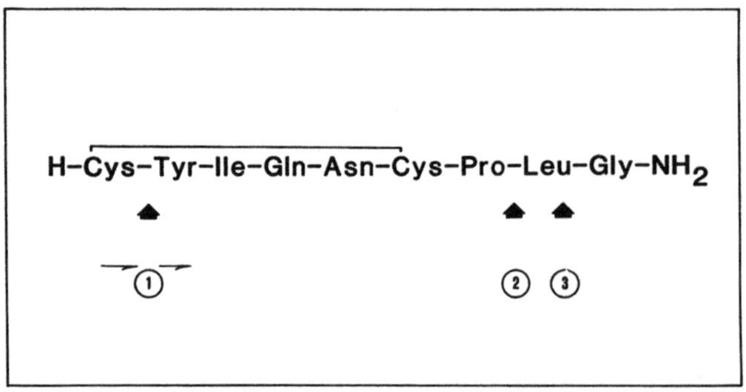

Fig. 2. Identified sites of initial proteolytic cleavage of oxytocin by brain peptidases. *1* aminopeptidase; *2* post-proline-cleaving enzyme; *3* carboxamido-peptidase or chymotrypsin-like peptidase

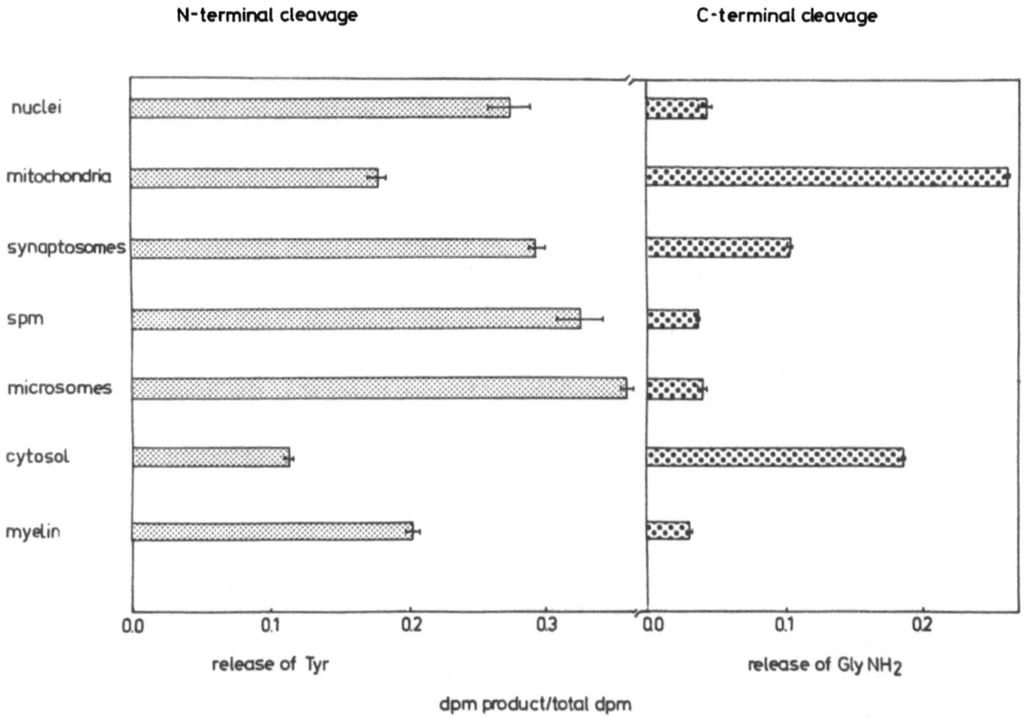

Fig. 3. Subcellular distribution of peptidase activities cleaving the N- and C-terminal portions of oxytocin. N-terminal-cleaving activity was quantitated by measuring the amount of [^{14}C]Tyr released from [^{14}C-Tyr^2]oxytocin-(1-9); C-terminal-cleaving activity is based on the release of [^{14}C]GlyNH$_2$ from [^{14}C-GlyNH$_2^9$]oxytocin-(1-9). (Adapted from Burbach et al. 1980a).
spm; synaptosomal plasma membranes, *dpm;* desintegrations per minute

minal dipeptide H-Leu-Gly-NH$_2$ (Walter 1972; Marks et al. 1973). Formation of GlyNH$_2$ was predominant in all subcellular fractions tested; H-Leu-Gly-NH$_2$ was found mainly in the microsomal and supernatant fractions. Walter et al. (1973) observed release of [^{14}C]GlyNH$_2$ by soluble and particulate preparations of rabbit and rat hypothalamus comparable to those used by Hooper (1963, 1964, 1966a, b, 1968). In addition, transient formation of H-Leu-Gly-NH$_2$ occurred. Initially, the amounts of H-Leu-Gly-NH$_2$ exceeded those of GlyNH$_2$, indicating that at least part of GlyNH$_2$ resulted from exopeptidase action on H-Leu-Gly-NH$_2$.

An enzyme able to cleave the Pro-Leu bond (post-proline-cleaving enzyme) has since been recognized (Schlank and Walter 1972; Koida and Walter 1976). The data showed that the C-terminal of oxytocin is susceptible to proteolysis by brain peptidases, and pointed to the Pro-Leu bond as the site of initial cleavage (Fig. 2). The subcellular distribution of peptidases acting on the C-terminal of oxytocin has been studied by determining the release of [^{14}C]GlyNH$_2$ from [^{14}C-GlyNH$_2^9$]oxytocin (Walter 1972; Marks et al. 1973; Burbach et al. 1980a). C-terminal-cleaving activity is by far highest in the cytosolic and mitochondrial fractions (Fig. 3), but was found to be low in nuclei, synaptosomal membranes, microsomes, and myelin preparations (Burbach et al. 1980a; Burbach 1980).

The existence of brain peptidases which attack the N-terminal of oxytocin was suggested by Walter et al. (Celis et al. 1971; Walter 1972; Walter et al. 1973), who studied the formation of an MSH-release-inhibiting factor (see also Sect. 2.5). Following the release of amino acids from oxytocin, the N-terminal cleavage was demonstrated in brain membranes (Fig. 3) (Burbach et al. 1980a). In the membrane preparation, N-terminal cleavages predominated over C-terminal cleavages; the release of free Tyr was about five times more rapid than the release of $GlyNH_2$ (Fig. 10). In preparations containing synaptosomes, synaptic membranes, microsomes and myeline, N-terminal-cleaving peptidase activity was relatively highest (Fig. 3).

From the above in vitro experiments it must be concluded that the N- and C-terminals of oxytocin are the sites of initial proteolytic attack by brain peptidases (Fig. 2). Cleavage of internal peptide bonds in the midportion of oxytocin has not been reported. Probably the rigid ring structure of this part of the molecule (Deslaurier and Smith 1970; Walter 1977; Smith 1981) renders the midportion inaccessible to peptidases. N- and C-terminal peptidases differ markedly in subcellular localization: synpatic and microsomal membranes contain mostly the N-terminal-cleaving peptidase activity; in the soluble and mitochondrial fractions C-terminal-cleaving activities are most abundant. Therefore, depending on the localization of oxytocin-cleaving peptidases and the availability of oxytocin to these sites, different types of metabolites can be expected to be generated from oxytocin.

2.3 Identification of Oxytocin Metabolites and Mechanism of Conversion

In view of the postulated role of oxytocin as a precursor of active fragments (Walter 1972; De Wied and Bohus 1978; Burbach and De Wied 1981), experiments have been focussed on the identification of products of the proteolysis of oxytocin by brain synaptic membranes. Formation of ^{14}C-labeled fragments in digests of [^{14}C-$GlyNH_2^9$]oxytocin with synaptic membranes was noted (Burbach et al. 1980a). These fragments accumulated concommitantly with the release of Tyr^2, indicating that they were formed by proteolysis of oxytocin at the N-terminal. However, the separation system (high-voltage paper electrophoresis) did not allow resolution of the components. HPLC analysis showed heterogeneity of the fragments (Burbach et al. 1980a). A preparative approach resulted in isolation of two components from digests of oxytocin, and preliminary characterization indicated that they were C-terminal fragments having the 4-9 sequence of oxytocin (Burbach et al. 1980c).

In more elaborate experiments, peptides that accumulated during exposure of oxytocin to synaptic membranes were fractionated and eventually isolated by HPLC (Burbach and Lebouille 1983). The peptides were chemically characterized by amino acid composition and N-terminal residues. A set of related C-terminal oxytocin fragments was identified; it included [Cyt^6]oxytocin-(2-9), -(3-9), -(4-9), -(5-9), and [$pGlu^4$,Cyt^6]oxytocin-(4-9) (Figs. 4 and 5). The accumulation of the

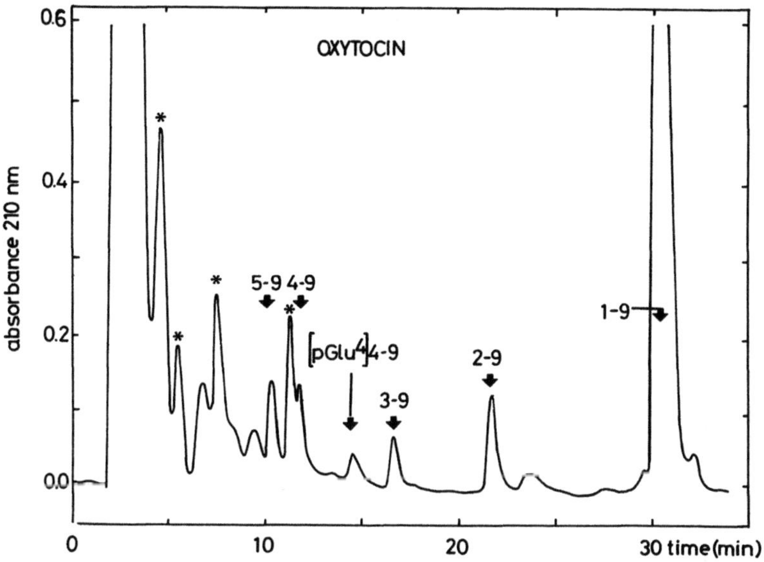

Fig. 4. Reverse-phase HPLC fractionation of oxytocin metabolites formed by in vitro incubation of oxytocin with brain membranes. The *arrows* indicate the position of identified fragments in the UV absorbance profile. [Cyt^6]oxytocin-(4-9) and -(5-9) were obtained after further purification from the indicated fraction.
* components derived from the membrane preparation. (Adapted from Burbach and Lebouille 1983)

latter peptide was favored under the in vitro incubation conditions due to cyclization of the N-terminal Gln into pGlu. When oxytocin was incubated at a higher concentration for a prolonged period, [pGlu^4,Cyt^6]oxytocin-(4-9) accumulated as the major peptide fragment (Burbach et al. 1983a).

The nature of oxytocin fragments identified in digests of oxytocin by brain synaptic membranes indicates that initial cleavage takes place at the Cys^1-Tyr^2 bond, and that proteolysis sequentially removes N-terminal amino acids. More direct evidence for such an aminopeptidase cleavage pattern was obtained from time course studies (Burbach and Lebouille 1983). Upon incubation of oxytocin-(1-9), [Cyt^6]oxytocin-(2-9) is the first peptide to appear; subsequently [Cyt^6]oxytocin-(3-9), -(4-9), [pGlu^4,Cyt^6]-(4-9), and [Cyt^6]oxytocin-(5-9) accumulate (Fig. 5). The conversion of oxytocin is accompanied by rapid release of free Tyr (50 n*M*/min), while the release of $GlyNH_2$ is slow (10 n*M*/min) (Fig. 10). This points to the predominance of the aminopeptidase cleavage mechanism. The pathway of conversion by this mechanism is depicted in Fig. 8.

The aminopeptidase-like mechanism of oxytocin cleavage by brain membranes resembles the cleavage of oxytocin by "plasma oxytocinase," an aminopeptidaselike enzyme circulating in the blood of pregnant women (Ferrier and Branda 1966; Sjöholm and Yman 1967). The mechanism of proteolysis of oxytocin is identical to that of vasopressin and vasotocin (see Sect. 3.1).

The cyclization of the N-terminal Gln^4 residue in the peptide [Cyt^6]oxytocin-(4-9) to pGlu is a modification with consequences for the pathway of conversion.

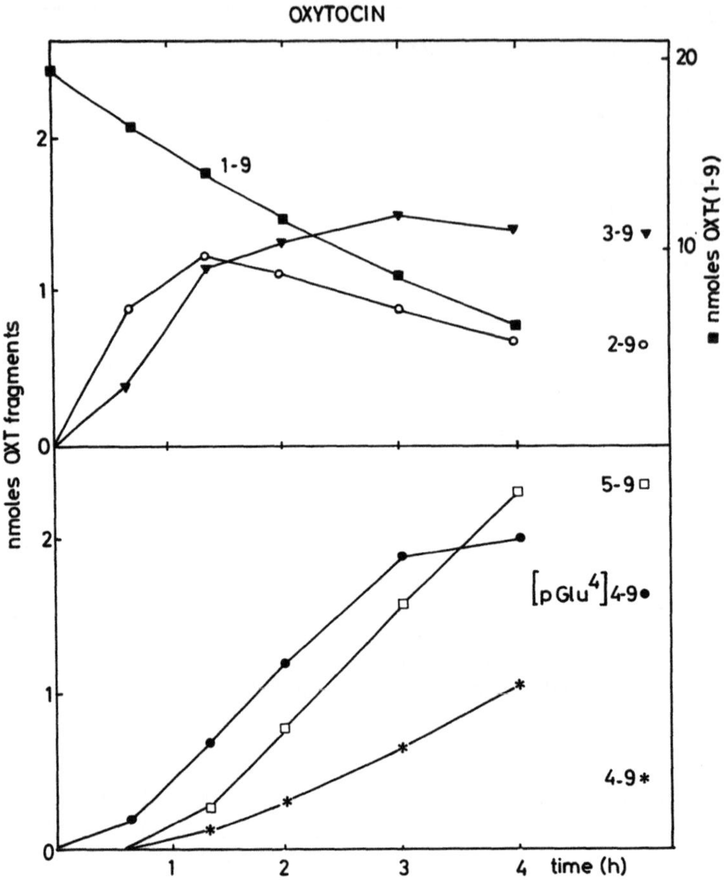

Fig. 5. Time course of conversion of oxytocin (OXT) and formation of metabolites. [^{14}C-GlyNH$_2^9$]oxytocin-(1-9) was incubated with brain synaptic membranes, and [^{14}C] peptides were separated by HPLC

The pGlu^4 residue is not removed by the aminopeptidase activity and serves as a block. The marked accumulation of [pGlu^4,Cyt^6]oxytocin-(4-9) during prolonged incubation of oxytocin (Burbach et al. 1983a) can be explained by the pGlu^4 formation. It is likely that pGlu^4 formation occurs spontaneously under the in vitro conditions, as is the case with other peptides (Folkers et al. 1970; Carraway and Leeman 1975; Abraham and Podell 1981).

A noted feature of the oxytocin fragments is the conserved disulfide bridge between Cys^1 and Cys^6, which shows that the aminopeptidase activity cleaves oxytocin without prior reduction of the disulfide bridge. This distinguishes the aminopeptidase from arylamidases purified from pig brain that were active in the presence of sulfhydryl-reducing agents and that acted more rapidly on S-benzylated nonapeptides than on intact oxytocin (Marks et al. 1973; Walter and Simmons 1977). Other reported purified brain aminopeptidases are inactive on oxytocin (Marks et al. 1973; Simmons and Brecher 1973).

Data on the properties of the oxytocin-cleaving aminopeptidase activity in brain membranes are still very scarce. The activity had a narrow pH optimum of pH 7.0 and its K_m was 6×10^{-5} *M* (Burbach et al. 1980a). Its action on [^{14}C-Tyr^2]oxytocin was most effectively inhibited by oxytocin-(1-9). The fragments oxytocin-(1-8) and -(1-7) had diminished inhibiting properties, while H-Pro-Leu-Gly-NH_2 and H-Leu-Gly-NH_2 did not inhibit the release of [^{14}C]Tyr (Burbach and De Wied 1981; J. P. H. Burbach, unpublished). The experiments may indicate specificity of the aminopeptidase activity for the intact nonapeptide structure. Support for aminopeptidase processing of oxytocin in vitro comes from pharmacologic work using aminopeptidase inhibitors. Meisenberg and Simmons (1984) reported that the behavioral excitation in mice induced by intracerebroventricular injection of oxytocin was potentiated by the aminopeptidase inhibitor amastatin. However, the duration of the effects of oxytocin analogues designed to be resistant to aminopeptidases was not extended. Instead, prolonged duration of behavioral action of deaminated vasopressin analogues was found. These approaches will be valuable for evaluation of the proteolytic mechanisms operative in situ in the brain.

2.4 Formation of H-Pro-Leu-Gly-NH_2 from Oxytocin

The tripeptide H-Pro-Leu-Gly-NH_2 has been considered as an MSH-release-inhibiting factor of the hypothalamus. It is the C-terminal tripeptide sequence of oxytocin (see Fig. 1). H-Pro-Leu-Gly-NH_2 has been isolated from hypothalamic tissue (Nair et al. 1971) and was reported to have intrinsic MSH-release-inhibiting activity (Celis et al. 1971). Incubation of oxytocin and lysine-vasopressin, but not N-terminal-protected oxytocin analogues, with median eminence microsomes and extracts led to the formation of MSH-release-inhibiting activity, as assessed by bioassay (Celis and Taleisnik 1971a; Celis et al. 1971; Walter 1974). The data suggested that H-Pro-Leu-Gly-NH_2 could be generated by aminopeptidase activity from oxytocin (Walter 1972, 1974). It was observed that incubation of [^{14}C-GlyNH_2^9]oxytocin-(1-9) with a hypothalamic particulate fraction of rabbit and rat resulted in the accumulation of a ^{14}C-radioactive component that comigrated with synthetic H-Pro-Leu-Gly-NH_2 upon high-voltage paper electrophoresis (Walter et al. 1973). However, no chemical characterization of this component was provided. The component accumulated slowly during prolonged incubation and represented less than 20% of the radioactivity at the longest time point. Walter et al. (1973) indicated that the component was H-Pro-Leu-Gly-NH_2, the proposed MSH-release inhibiting factor (Celis and Taleisnik 1971a; Nair et al. 1971; Celis et al. 1971), and suggested that it could have been formed by stepwise proteolysis of oxytocin from the N-terminal onward (Celis et al. 1971; Walter et al. 1973).

Later experiments failed to detect H-Pro-Leu-Gly-NH_2 as a product of proteolysis of [^{14}C-GlyNH_2^9]oxytocin by brain membranes (Burbach et al. 1980a; Burbach and Lebouille 1983; Burbach et al. 1983a) or a particulate rat hypothalamic fraction as used by Walter et al. (1973) (J. P. H. Burbach, unpublished). Instead, a number of larger C-terminal peptides were detected, ranging from

[Cyt^6]oxytocin-(2-9) to [Cyt^6]oxytocin-5-9) (see Sect. 2.3). The mechanism of their formation was by stepwise aminopeptidase conversion as indicated for H-Pro-Leu-Gly-NH_2 (Celis et al. 1971; Walter 1972, 1974), but the Cys^6-Pro^7 bond, a peptide bond likely to be split only with great difficulty by peptidases, was not cleaved. The endogenous presence of H-Pro-Leu-Gly-NH_2 in brain has been investigated, but the peptide could not be demonstrated in the hypothalamus (Kastin et al. 1981 a, b; Manberg et al. 1982).

In these experiments, which combined chromatographic characterization with radioimmunologic detection, other cross-reacting components were observed. Taken together, the currently available data do not exclude the possibility, but make it unlikely, that H-Pro-Leu-Gly-NH_2 is released enzymatically from oxytocin by hypothalamic peptidases and occurs endogenously. Other oxytocin fragments that are readily generated from oxytocin during exposure to brain membranes, for instance [pGlu^4,Cyt^6]oxytocin-(4-9), may be similar or closely related to the component described by Walter et al. (1973), and might represent some of the H-Pro-Leu-Gly-NH_2-like immunoreactivity detected in the brain (Kastin et al. 1981 a, b; Manberg et al. 1982). In addition, the MSH-release-inhibiting activity of H-Pro-Leu-Gly-NH_2 in the rat has been disputed (Grant et al. 1973; Thody et al. 1980). In the frog, H-Pro-Leu-Gly-NH_2 is inactive (Tonon et al. 1979). However, in preliminary experiments [pGlu^4,Cyt^6]oxytocin-(4-9) was also inactive in stimulating the release of α-MSH from the perfused frog neurointermediate pituitary (M. Lamacz, M.-C. Tonon and H. Vaudry, unpublished). The above experiments raise doubts about the existence, formation and biologic effects of H-Pro-Leu-Gly-NH_2. Some of the reported findings may be ascribed to other closely related oxytocin fragments, and deserve further attention.

2.5 Conversion of H-Pro-Leu-Gly-NH_2 by Brain Peptidases

A number of studies have been concerned with the degradation of H-Pro-Leu-Gly-NH_2. In view of the identity of this peptide with the C-terminal of oxytocin, the data provide additional information about peptidase action on the C-terminal of oxytocin. They are briefly reviewed below.

In vitro degradation of H-Pro-Leu-Gly-NH_2 in rat brain extracts occurs by initial cleavage of the Pro-Leu bond (Marks and Walter 1972; Marks et al. 1974). The same mechanism has been indicated by Hui et al. (1980), on the basis of the release of amino acids liberated during in vitro incubation of H-Pro-Leu-Gly-NH_2. De Wildt et al. (1982) used H-Pro-[^{3}H]Leu-Gly-NH_2 as the substrate in in vitro experiments. [^{3}H]Leu was rapidly released, and H-[^{3}H]Leu-Gly-NH_2 was the only detectable intermediate. The amounts were very low, indicating that H-Leu-Gly-NH_2 is only a very short-lived metabolite. H-[^{3}H]Leu-Gly-NH_2 was also found in the brain in small amounts after intravenous and intracerebroventricular administration of H-Pro-[^{3}H]Leu-Gly-NH_2 (De Wildt et al. 1982), indicating that the Pro-Leu cleavage is also operative in vitro.

Several brain peptidases have been indicated as candidates to split the Pro-Leu bond of H-Pro-Leu-Gly-NH_2. Marks and Walter (1972) purified a peptidase, arylamidase B, which cleaved this bond. This peptidase activity might be similar to the "MIF aminopeptidase" described by Hui et al. (1980). The arylamidase B activity does not split the Pro-Leu bond of oxytocin, nor does it act on the N-terminal of oxytocin without disulfide reduction (Marks et al. 1973). It is of interest to note that the post-proline-cleaving enzyme that cleaves Pro^7-Leu^8 in oxytocin does not act on H-Pro-Leu-Gly-NH_2 (Koida and Walter 1976). Also, other peptidases that hydrolyze peptide bonds in the C-terminal of oxytocin do not cleave H-Pro-Leu-Gly-NH_2 or H-Leu-Gly-NH_2 (Simmons and Walter 1980, 1981).

Despite the structural similarity of H-Pro-Leu-Gly-NH_2 and the C-terminal of oxytocin, it must be concluded from the above experiments that very little, if any, similarity exists between the peptidases acting on these two peptides. It might be that peptidases with completely different substrate specificities are involved in the proteolysis of these peptides.

2.6 Endocrine Influences on the Proteolysis of Oxytocin

Most work on endocrine influences on the proteolysis of oxytocin in the brain has been concerned with the effects of gonadal steroids. This line of research was initiated by Hooper (1964), from the viewpoint that oxytocin-inactivating peptidase activity might be related to some aspect of the control of oxytocin metabolism in the hypothalamus; the concentration of peptidase activity was suggested as an index of hormone production (Hooper 1966b). The loss of oxytocin bioactivity was used in these studies as a parameter for oxytocin inactivation. Kuhl, Taubert, and colleagues studied the effects of gonadal steroids on hypothalamic peptidases, using L-cystine-β-naphthylamide or L-cystine-p-nitroanilide as substrates (Tuppy et al. 1962; Kuhl et al. 1974a, b). The cleavage of these synthetic substrates resembles the cleavage of the N-terminal of oxytocin. Therefore, the peptidase activity measured, indicated as L-cystine arylamidase or cystine aminopeptidase, may overlap with the aminopeptidase which cleaves the N-terminal of oxytocin (Burbach and Lebouille 1983).

Hooper (1964) reported an increased oxytocin-inactivating activity in a particulate fraction of the dog hypothalamus during late pregnancy. In the rabbit, enzyme activity increased after mating and during early pregnancy, coinciding with the time of blastocyst implantation (Hooper 1966a; Frith and Hooper 1971a), after parturition, and during lactation (Hooper 1966b). Enzyme activities were reduced by ovariectomy and stimulated by in vivo treatment with estradiol and synthetic ovulation inhibitors (Hooper 1963; Frith and Hooper 1971b). It was shown by in vitro incubation of hypothalamic slices that estrogens acted directly on the hypothalamus in stimulating oxytocin-inactivating activities. The stimulation was blocked by cycloheximide (Frith and Hooper 1974). In general, peptidase activities in both supernatant and particulate fractions responded to treatments, but behaved differently with respect to magnitude and time dependence (Griffith and Hooper 1973a–d).

In the female rat, the oxytocin-inactivating activity in the supernatant fraction, but not in the particulate preparation, decreased upon ovariectomy (Griffith and Hooper 1972a) and on the afternoon of proestrus (Griffith and Hooper

1973c). Neonatal administration of testosterone to female rats resulted in lower supernatant and particulate activities than those in male rats (Griffith and Hooper 1972b).

In a similar type of experiment, Kuhl, Taubert, and colleagues determined L-cystine arylamidase activity in hypothalamic homogenates, lyophilized extracts, and cytosolic fractions. They found that estrogens stimulated L-cystine arylamidase activity in male and female rats (Taubert et al. 1970; Heil et al. 1971; Bickel et al. 1972; Kuhl et al. 1974a, Kuhl and Taubert 1975b). The effects of testosterone and progesterone were sex-dependent (Bickel et al. 1972; Kuhl et al. 1974a, b). Testosterone stimulated L-cystine arylamidase activity in the male rat only. Also, influences of age per se and age-dependent responses to steroids were observed, and were generally sex-linked (Kuhl et al. 1974a, b).

Celis and Taleisnik studied influences of estrogens on the in vitro formation of MSH-release-inhibiting and -stimulating factors. These two factors were reportedly generated during digestion of oxytocin by a hypothalamic microsomal and mitochondrial preparation, respectively, as determined by bioassay (Celis and Taleisnik 1971a, b; Celis et al. 1971). The microsomal formation of the MSH-release-inhibiting factor was stimulated by estrogens (Celis and Taleisnik 1974). The mitochondrial formation of the MSH-release-stimulating factor was detected only during proestrus and estrus (Celis 1975). The data suggest that estrogens differentially influence the conversion of oxytocin in microsomal and mitochondrial fractions of the hypothalamus, and may relate to some of the findings of the group of Hooper with respect to the particulate oxytocin-inactivating enzyme activity.

The above experiments all point to a stimulating influence of estrogens on hypothalamic peptidases able to cleave oxytocin. However, the nature of these peptides and the site at which they attack the oxytocin molecule remain obscure. In Hooper's approach – using loss of oxytocic bioactivity as a measure for oxytocin cleavage – the total pool of oxytocin-cleaving peptidases was determined. Experiments of Walter and colleagues, using [^{14}C-GlyNH$_2^9$]oxytocin, suggested that both supernatant and particulate preparations contained mainly peptidases that cleaved the C-terminal of oxytocin (Walter 1972; Walter er al. 1973). In contrast, Burbach et al. (1980a), using [^{14}C-Tyr^2]- and [^{14}C-GlyNH$_2^9$]oxytocin, found primarily N-terminal cleavage of the molecule in the soluble fraction (see also Fig. 3). Kuhl and Taubert used cystine derivatives to assess peptidase activity (Tuppy et al. 1962). The cleavage of these substrates by the L-cystine arylamidase activity resembles the aminopeptidase-type cleavage of oxytocin.

It has been shown that the measured peptidase activity was not always entirely specific for oxytocin: the oxytocin-inactivating enzyme activity as determined by Hooper et al. was completely inhibited by luteinizing hormone-releasing hormone (LH-RH), but not by thyrotropin-releasing hormone (TRH) (Griffith and Hooper 1974). Oxytocin, lysine-vasopressin, LH-RH and TRH competitively inhibited the L-cystine arylamidase activity (Kuhl und Taubert 1975a). Therefore, it may be assumed that the data only partly relate to the peptidases that are functionally involved in the hypothalamic metabolism of oxytocin. Several of the findings have been interpreted as being associated with the release of LH from the pituitary gland (Griffith and Hooper 1973b, c, d) and with the metabolism of LH-

Table 1. Oxytocin-converting peptidase activities in the hypothalamus of the Brattleboro rat, determined in a crude membrane preparation by separate incubation of [^{14}C-Tyr^{2}] oxytocin-(1–9) and [^{14}C-GlyNH$_2$^{9}]oxytocin-(1–9). Free ^{14}C-amino acids and peptides were separated by HPLC[a]

Animal	Aminopeptidase [^{14}C]Tyr^{2} release (%)	C-terminal-cleavage [^{14}C]GlyNH$_2$^{9} release (%)
HONO	40.4±3.6	38.7±10.3
HODI	41.2±3.4	43.3± 8.2

[a] Values represent the percentage of free ^{14}C-amino acid of total radioactivity (±SD) obtained from homozygous normal (HONO) and homozygous diabetes insipidus (HODI) Brattleboro rats. (Data unpublished)

RH (Griffith et al. 1973; Griffith and Hooper 1974; Kuhl and Taubert 1975a, b; Kuhl et al. 1977, 1978). Further experiments using more specific methods to determine oxytocin-converting peptidases or employing biosynthetic techniques will be required to draw firm conclusions on the regulation of oxytocin conversion in the brain by endocrine factors.

The activity of peptidases cleaving the N- and C-terminals of oxytocin has been studied in the Brattleboro strain in order to investigate whether the absence of vasopressin influences the metabolism of oxytocin (J.P.H. Burbach, unpublished). Brattleboro rats with homozygous diabetes insipidus caused by a genetic defect in vasopressin synthesis (Schmale and Richter 1984) were compared to homozygous normal rats with respect to the conversion of oxytocin by a membrane preparation from the limbic brain. The release of [^{14}C]Tyr and [^{14}C]GlyNH$_2$ was taken as the index of N- and C-terminal-cleaving peptidases respectively. As is shown in Table 1, no statistical difference between the peptidase activities was observed in the two groups. Apparently, the lack of vasopressin has no significant effect on the activity of the oxytocin-converting peptidases.

3 Proteolytic Conversion of Vasopressin and Vasotocin

Vasopressin and vasotocin are structurally related to oxytocin (Fig. 1). As a hormone, vasopressin has activities independent from those of oxytocin. As neuropeptides in the brain, vasopressin and oxytocin often display opposite activities (De Wied 1980, 1983). Vasotocin is the nonmamalian nonapeptide. In some species it might possess both oxytocinergic and vasopressinergic activities (Acher 1981; Sawyer 1964, 1977).

3.1 Identification of Metabolites and Pathway of Conversion

The first observations of the inactivation of vasopressin by brain tissue were made as early as 1935 by Heller and Urban (1935). They observed that "pituitrin," a posterior lobe extract with antidiuretic activity, lost this activity when incubated with minced guinea pig brain. They explained this phenomenon as absorption into the tissue suspension, but also concluded that "a quickly acting ferment destroys the pituitrin ..." (Heller and Urban 1935).

Evidence of enzymatic cleavage of vasopressin in the brain and characterization of products was produced by Pliška et al. (1971 b). Using [^{3}H-Tyr^{2},Lys^{8}]-vasopressin, these authors reported that the peptide was rapidly taken up by minced brain cortex and neurohypophysis. After separation of radioactive components by thin-layer chromatography (Pliška et al. 1971 a), free [^{3}H]Tyr and an aminopeptidase split-product were detected (Pliška et al. 1971 b). The results from the two types of tissue were comparable and pointed to initial cleavage of [Lys^{8}]vasopressin at the Cys^{1}-Tyr^{2} bond. In addition, the existence of a C-terminal cleavage at the Lys^{8}-GlyNH$_2^9$ bond was indicated. Marks et al. (1973) reported that [^{14}C]GlyNH$_2$ was released from [^{14}C-GlyNH$_2^9$,Arg^{8}]vasopressin by a brain supernatant fraction; only traces of the dipeptide H-Arg-Gly-NH$_2$ were found. Both N- and C-terminal cleavages can be observed when [Arg^{8}]vasopres-

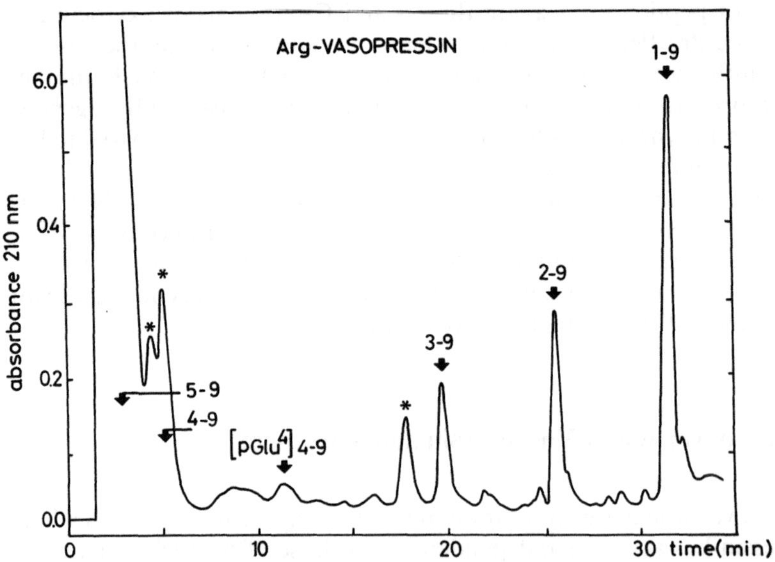

Fig. 6. Reverse-phase HPLC fractionation of [Arg^{8}]vasopressin metabolites formed by in vitro incubation of [Arg^{8}]vasopressin-(1-9) with brain membranes. The *arrows* indicate the position of identified fragments in the UV absorbance profile. The fragments [Cyt^{6}, Arg^{8}]vasopressin-(2-9), -(3-9) and [pGlu^{4},Cyt^{6},Arg^{8}]vasopressin-(4-9) were obtained in pure form from this HPLC system. The peptides [Cyt^{6},Arg^{8}]vasopressin-(4-9) and -(5-9) were isolated from the indicated fractions by a subsequent ion-paired reverse-phase HPLC step.

* components derived from the membrane preparation. (Unpublished)

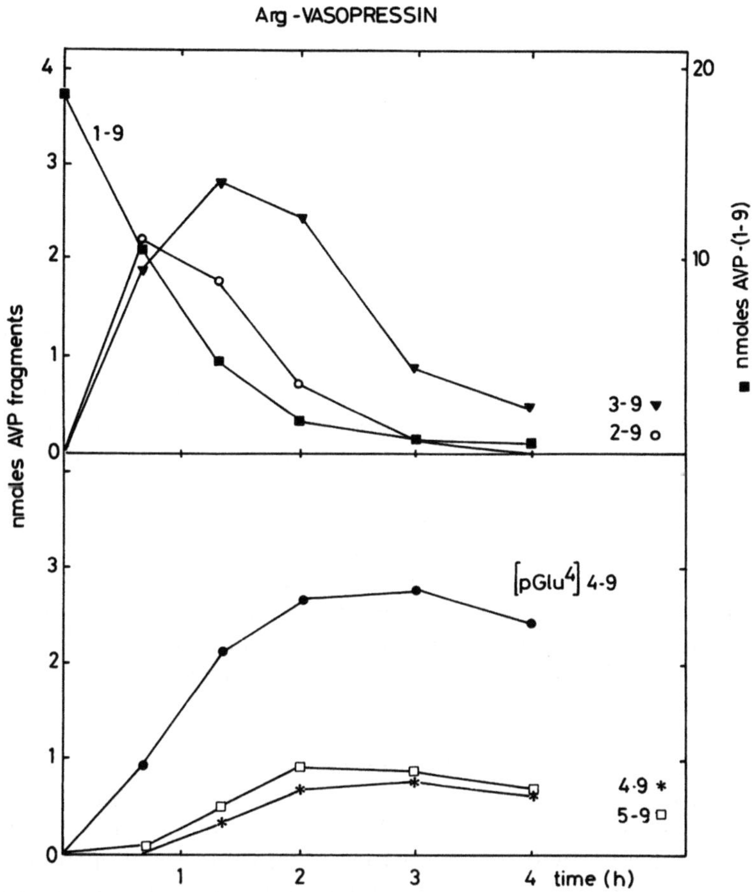

Fig. 7. Time course of conversion of [Arg^8]vasopressin (AVP) and formation of metabolites. [Arg^8,^{14}C-GlyNH$_2^9$]vasopressin-(1-9) was incubated with a rat brain membrane preparation and [^{14}C]fragments were separated by reverse-phase HPLC

sin is exposed to brain synaptic membrane preparations (Fig. 10). However, the N-terminal cleavages are predominant; C-terminal cleavages that result in the liberation of [^{14}C]GlyNH$_2^9$ from [^{14}C-GlyNH$_2^9$,Arg^8]vasopressin constitute only 10% of the N-terminal events leading to the liberation of [^{14}C]Tyr from [^{14}C-Tyr^2,Arg^8]vasopressin (Burbach and Lebouille 1983; Fig. 10).

Upon incubation with this membrane preparation, a series of [Arg^8]vasopressin fragments are formed (Fig. 6). These have been isolated by HPLC and chemically characterized. They all represent aminopeptidase split-products. The precursor-product relationship of these fragments, as suggested by their structure, is demonstrated by time course experiments (Fig. 7). The initial product is the open nonapeptide [Cyt^6,Arg^8]vasopressin-(2-9), which is further converted into [Cyt^6,Arg^8]vasopressin-(3-9). Subsequently, [Cyt^6,Arg^8]vasopressin-(4-9), -(5-9), [pGlu^4,Cyt^6,Arg^8]vasopressin-(4-9), and a fraction tentatively identified as containing [Cyt^6,Arg^8]vasopressin-(6-9) (J.P.H. Burbach and X.C. Wang, unpublished) are formed in a time-dependent fashion (Fig. 7; Burbach and Lebouille 1983). This proteolytic mechanism is indicative of the involvement of an aminopeptidase and is similar to that of the conversion of oxytocin; it is summarized in Fig. 8. The aminopeptidase pathway is interrupted by cyclization of the N-terminal Gln^4 to $pGlu^4$ (Fig. 8). The resulting peptide, [$pGlu^4$,Cyt^6,Arg^8]vasopressin-(4-9), is not cleaved by the aminopeptidase activity and accumulates in relatively large amounts during prolonged incubation of [Arg^8] vasopressin (Burbach et al. 1983b). When [Cyt^6,Arg^8]vasopressin-(3-9) was used as the substrate, [$pGlu^4$,Cyt^6,Arg^8]vasopressin-(4-9) was also rapidly formed (Wang and Burbach 1986). The significance of this fragment as a peptide with potent and selective central activity is discussed in Sect. 4.2.

Observations of Constantini and Pearlmutter (1984) also point to the aminopeptidase pathway. These authors used [^3H-Phe^3,Arg^8]vasopressin to study the stability and degradation of vasopressin during in vitro experiments. Radioactive metabolites were observed which could well be aminopeptidase split-products. Interestingly, the proteolysis of vasopressin was inhibited by 1,10-phenanthroline and high concentrations of benzamidine, Zn^{2+}, Ni^{2+} and Co^{2+}. The serine protease inhibitor, phenylmethylsulfonyl fluoride, was inactive, as was leupeptide in low concentration. It was suggested that the proteolytic activity cleaving vasopressin was a metallopeptidase (Constantini and Pearlmutter 1984). The aminopeptidase inhibitors amastatin, bestatin, and bacitracin were ineffective in inhibiting the degradation of vasopressin.

Further classification of a vasopressin-converting aminopeptidase activity was obtained recently (J.P.H Burbach, D. Terwel and J.L.M. Lebouille, unpublished). Using [^3H-Phe^3,Arg^8]vasopressin-(1-9) as substrate, rat brain membranes as source of enzyme activity and hydrophobic chromatography to assay the aminopeptidase activity sensitivity to both thiol and metallopeptidase inhibitors was observed (Table 2). Parachloromercury benzoate (PCMB), Hg^{2+} and EDTA were effective inhibitors of the vasopressin-converting aminopeptidase activity. Enzyme activity was stimulated by low concentrations of β-mercaptoethanol and dithiothreitol. Dependence of enzyme activity on a metal ion was demonstrated by 1) inhibition by the metalchelator EDTA (Table 2) and 2) full recovery of EDTA-treated enzyme activity by 10 μM Zn^{2+}. The vasopressin-converting aminopeptidase activity had a pH optimum at pH 7.0, and a K_m of 12×10^{-6} M.

Vasotocin ([Arg^8]vasotocin-(1-9); Fig. 1) is subject to the same proteolytic mechanisms as oxytocin and [Arg^8]vasopressin, as reported by Wang et al. (1983a). A series of related C-terminal fragments, ranging from [Cyt^6]vasotocin-(2-9) to [Cyt^6]vasotocin-(5-9), were characterized as products of vasotocin conversion by peptidases in rat and chicken brain membranes (Fig. 9). Although the nature of generated fragments was identical for both species, differences in conversion rates were noted. Under the in vitro conditions, the rate of vasotocin conversion by chicken brain membranes was about twice as slow as by rat brain membranes. The possible significance of vasotocin fragments is discussed in Sect. 4.2.

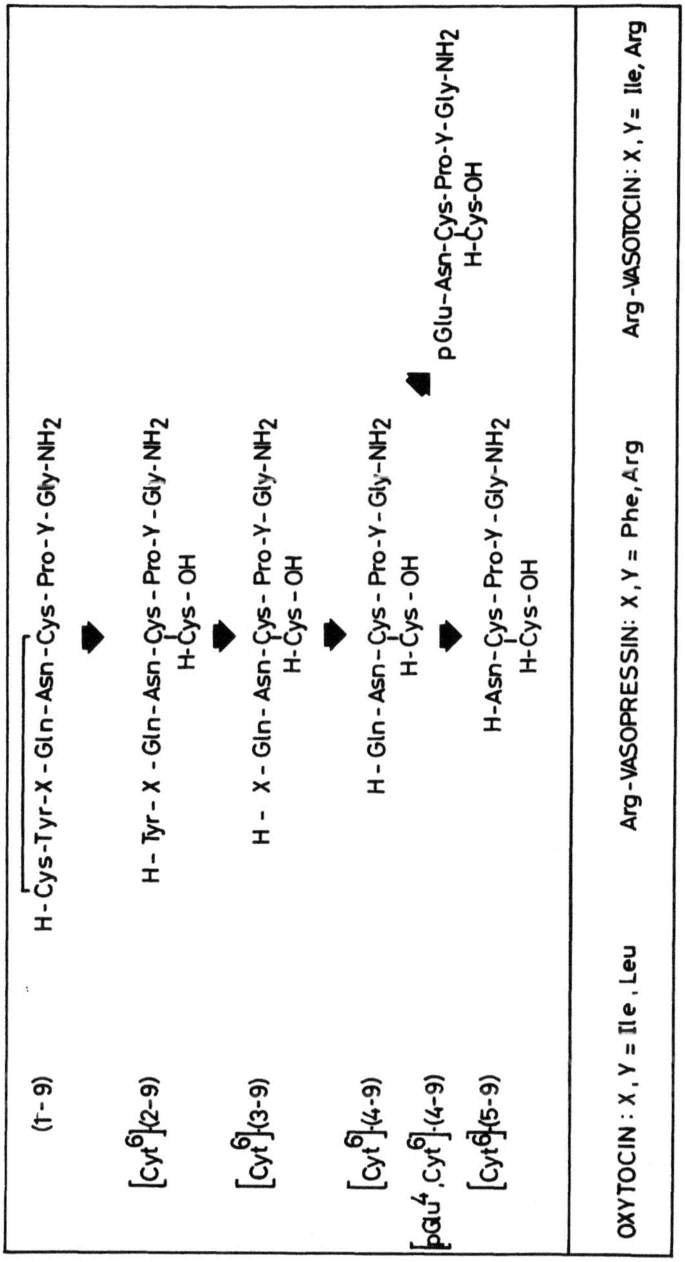

Fig. 8. Aminopeptidase pathway of conversion of oxytocin, [Arg⁸]vasopressin, and vasotocin. The sequence of proteolytic events and the structure of products are shown. The pathway is based on the identification of peptide fragments and time course studies performed by in vitro incubation of the nonapeptides with rat brain membrane preparations

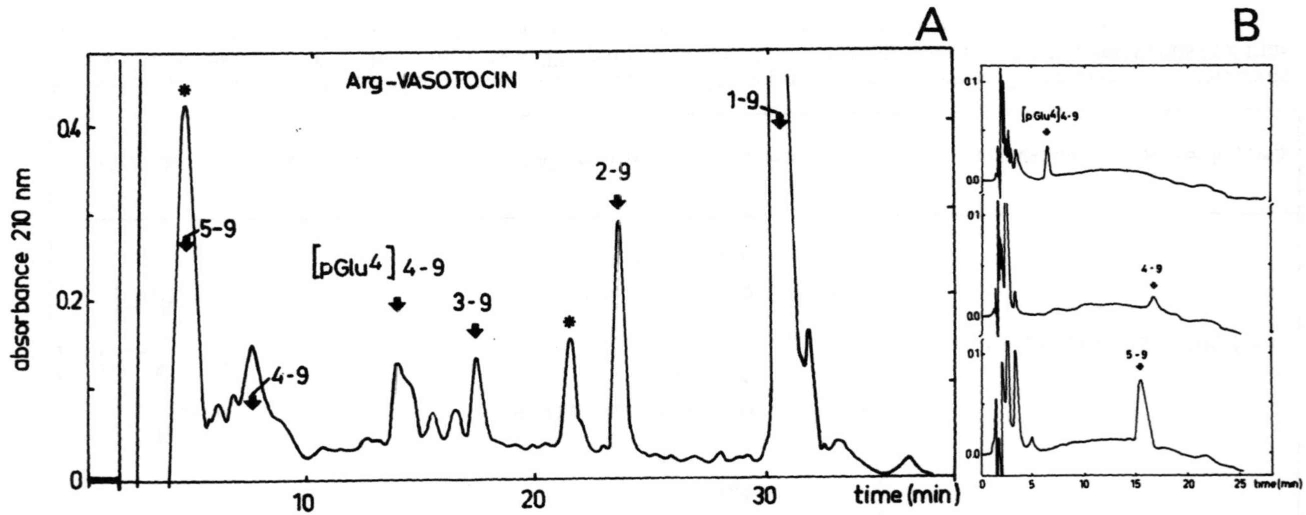

Fig. 9 A, B. Isolation of vasotocin metabolites accumulated during incubation of [Arg^8]vasotocin-(1-9) with rat brain membranes. **A**, reverse-phase HPLC of vasotocin fragments: [Cyt^6]vasotocin-(2-9) and -(3-9) were obtained in pure form; **B**, purification of [pGlu^4,Cyt^6]vasotocin-(4-9); [Cyt^6]vasotocin-(4-9) and -(5-9) from the fractions shown in **A** by ion-paired reverse-phase HPLC.
* components derived from the membrane preparation. (Adapted from Wang et al. 1983 a)

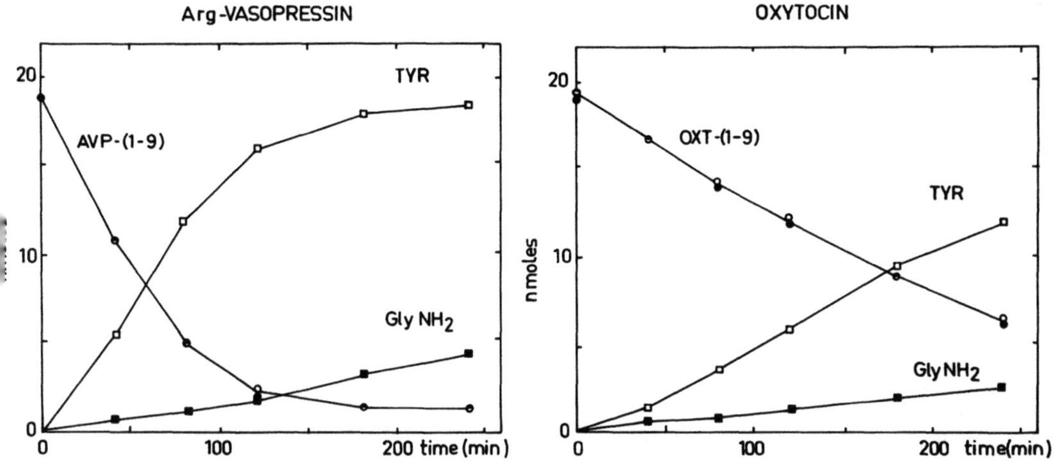

Fig. 10. Comparison of the conversion of oxytocin and [Arg^8]vasopressin by brain membranes in vitro. The disappearance of intact peptide and the release of free Tyr and $GlyNH_2$ are shown. The release of Tyr reflects the N-terminal cleavage of oxytocin and [Arg^8]vasopressin; $GlyNH_2$ is derived from C-terminal cleavages. Peptides and amino acids were determined by HPLC fractionation and ^{14}C-scintillation counting. (From Burbach et al. 1982)

3.2 Comparison of the Proteolysis of Oxytocin, Vasopressin, Vasotocin, and Analogues

Oxytocin, [Arg^8]vasopressin, and vasotocin are similarly converted in vitro by an aminopeptidaselike enzyme in brain membranes to a number of related C-terminal fragments (Burbach and Lebouille 1983; Wang et al. 1983a). These observations and the similarity in N-terminal structure of the nonapeptides suggest that the same aminopeptidase activity can act on the three substrates. This suggestion is supported by experiments showing inhibition of the aminopeptidase activity acting on [^{14}C-Tyr^2]oxytocin by [Arg^8]vasopressin, [Lys^8]vasopressin, and vasotocin (J.P.H. Burbach, unpublished). Although the mechanism of conversion and, probably, the aminopeptidase activity involved are similar, there are marked differences in susceptibility of the peptides to this type of proteolysis (Burbach and Lebouille 1983; Burbach et al. 1982). Under certain in vitro conditions (Fig. 10), oxytocin is converted with a half-life of 200 min; [^{14}C]Tyr is released by the aminopeptidase as a rate of approximately 50 nM/min. In this situation, [Arg^8]vasopressin has a half-life of 40 min, and [^{14}C]Tyr appears at 140 nM/min. The accumulation of [^{14}C]$GlyNH_2$, mediated by C-terminal cleavages, occurs at a rate of 10 nM/min and 13 nM/min respectively. The data indicate that the difference in the conversion rates of oxytocin and [Arg^8]vasopressin is due to a different susceptibility of the peptides to the aminopeptidase activity.

Comparison of the conversion of a number of analogues in vitro showed that all peptides with Ile^3 (oxytocin, vasotocin, and tocinamide) were more resistant to peptidase action than those having Phe^3 ([Arg^8]vasopressin, oxypressin, and

Table 2. Inhibition of vasopressin converting aminopeptidase activity by various substances

Substance	Concentration	Activity (%)
None	–	100
N-ethylmaleimide (N-EM)	0.1 m*M*	50
p-chloromercury benzoate (PCMB)	0.1 m*M*	22
Dithiothreitol (DTT)	0.1 m*M*	261
$MgCl_2$	0.1 m*M*	20
$ZnCl_2$	0.1 m*M*	6
EDTA	0.1 m*M*	19
Phenylmethanesulfonyl fluoride (PMSF)	1 m*M*	86
Aprotinin	100 $\mu g \cdot ml^{-1}$	100
Leupeptin	100 $\mu g \cdot ml^{-1}$	84
Antipain	100 $\mu g \cdot ml^{-1}$	50
Bacitracin	100 $\mu g \cdot ml^{-1}$	40

pressinamide, Table 3; Burbach et al. 1982). Thus, the residue at position 3 of the peptides – Ile or Phe (Fig. 1) – determines primarily the susceptibility to aminopeptidase cleavage, probably due to conformational influences (Walter 1977; Smith 1981). In addition, small influences of the C-terminal tail on the N-terminal cleavage can be seen (Table 3). The difference in proteolysis of oxytocin and [Arg^8]vasopressin by brain peptidases can have consequences for the profile of active fragments formed from these peptides.

A difference in the proteolysis of oxytocin and [Arg^8]vasopressin has also been observed in a peripheral tissue (kidney), but is of an entirely different nature (Walter and Schlank 1975). [^{14}C]$GlyNH_2$-labeled oxytocin and [Arg^8]vasopressin were degraded at similar rates by kidney homogenates, but the accumulation of [^{14}C]$GlyNH_2$ was very rapid in the case of oxytocin and initially slow for [Arg^8]vasopressin. There was very little formation of the C-terminal dipeptide H-Leu-Gly-NH_2 from oxytocin, while the accumulation of H-Arg-Gly-NH_2 from [Arg^8]vasopressin was considerable. The difference has been explained by the presence of an enzyme that cleaves the Leu-$GlyNH_2$ bond, but not the Arg-$GlyNH_2$ bond. Such a chymotrypsinlike peptidase has been characterized by Koida et al. (1971).

3.3 [Arg^8]Vasopressin Metabolites in the Brain

Based on the observation that [Arg^8]vasopressin metabolism in vitro generates C-terminal fragments with central activity (see Sect. 4.2), a preliminary investigation of the endogenous existence of such metabolites in the brain has been conducted (Burbach et al. 1984b). Experiments aimed at investigating their presence in the brain have employed immunologic and chromatographic techniques. Dissected rat brain areas were extracted and the extract chromatographed by HPLC. By a

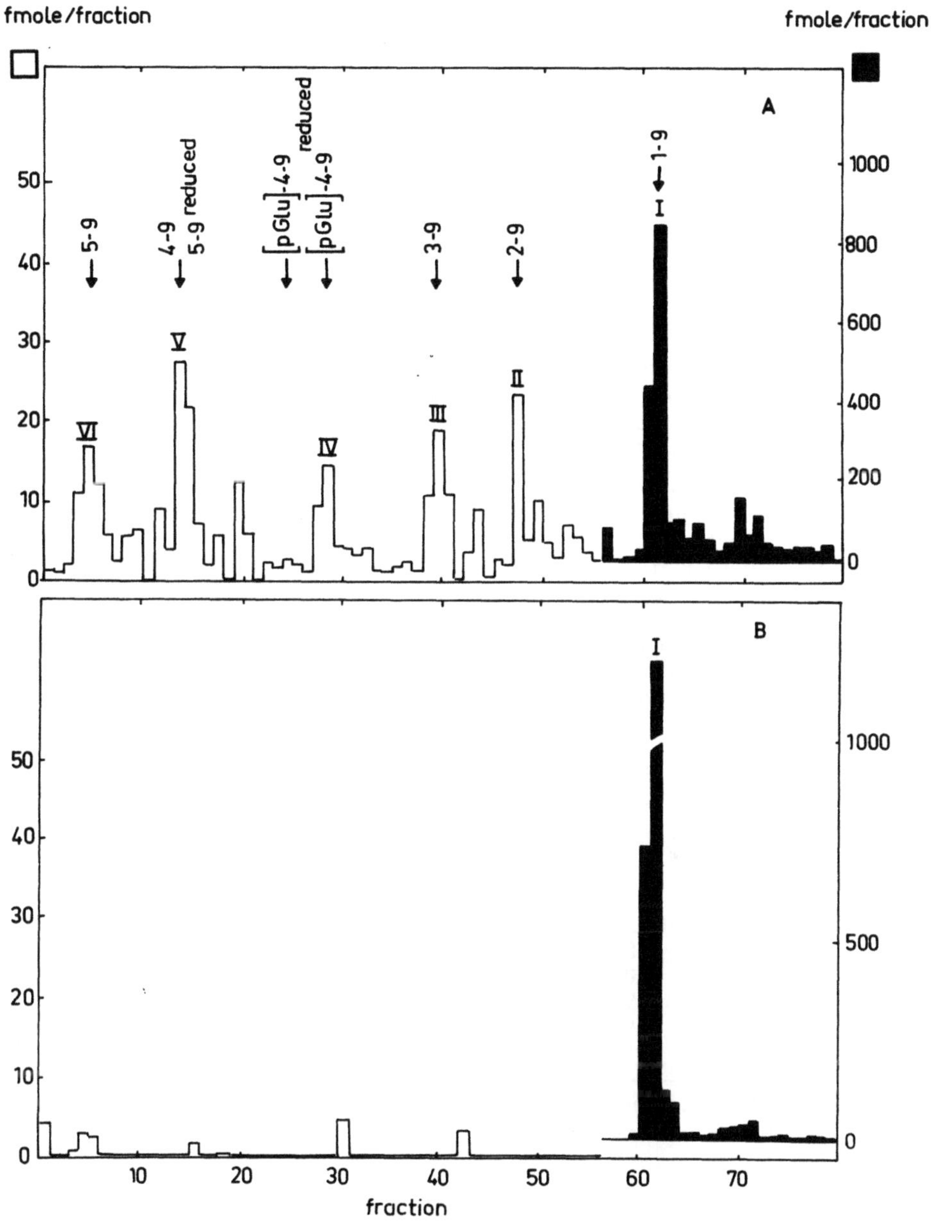

Fig. 11 A, B. Endogenous C-terminal [Arg[8]]vasopressin immunoreactive peptides in the rat hippocampus. Hippocampal tissue was extracted and fractionated by reverse-phase HPLC. Fractions were subjected to radioimmunoassay using an antiserum directed to the C-terminal of [Arg[8]]vasopressin **A** or an antiserum requiring the N-terminal portion of [Arg[8]]vasopressin **B**. The *arrows* indicate the elution position of synthetic [Arg[8]]vasopressin fragments determined in a parallel run

Table 3. Comparison of resistance of oxytocin, vasopressin, and derivatives to peptidases in rat brain membranes

Peptide	Structure[a]	Degradation rate[b]
Oxytocin	*H*-Cys-Tyr-*Ile*-Gln-Asn-Cys-Pro-*Leu*-GlyNH$_2$	4.4
[Arg^8]vasotocin	*H*-Cys-Tyr-*Ile*-Gln-Asn-Cys-Pro-*Arg*-GlyNH$_2$	7.2
Tocinamide	*H*-Cys-Tyr-*Ile*-Gln-Asn-Cys-NH$_2$	5.2
Oxypressin	*H*-Cys-Tyr-*Phe*-Gln-Asn-Cys-Pro-*Leu*-GlyNH$_2$	11.4
[Arg^8]vasopressin	*H*-Cys-Tyr-*Phe*-Gln-Asn-Cys-Pro-*Arg*-GlyNH$_2$	11.6
[Lys^8]vasopressin	*H*-Cys-Tyr-*Phe*-Gln-Asn-Cys-Pro-*Lys*-GlyNH$_2$	11.5
Pressinamide	*H*-Cys-Tyr-*Phe*-Gln-Asn-Cys-NH$_2$	16.2
Des-amino-[Arg^8]-vasopressin	Cys-Tyr-*Phe*-Gln-Asn-Cys-Pro-*DArg*-GlyNH$_2$	0.5

[a] Positions of structural difference are shown in italics

[b] Degradation rates were calculated as rate constants of zero-order enzyme reaction. (Adapted from J.P.H. Burbach et al. 1982, unpublished)

Table 4. Content of C-terminal [Arg^8]vasopressin fragments in comparison with [Arg^8]-vasopressin (AVP) in the rat hypothalamus, amygdala, and pituitary gland. Components in brain extracts were separated by HPLC and subsequently quantitated by radioimmunoassay, using a C-terminal specific antiserum. (Adapted from J.P.H. Burbach et al. 1984b, unpublished)[a]

Component	Hypothalamus		Hippocampus		Amygdala		Pituitary	
	pmol/mg protein	%	fmol/mg protein	%	fmol/mg protein	%	fmol/mg protein	%
I (AVP)	18.27	100	37.80	100	384.8	100	1913.4	100
II	0.23	1.80	1.66	4.40	39.5	10.3	11.59	0.61
III	0.25	2.00	3.61	9.60	52.4	13.6	1.65	0.09
IV	0.24	1.90	3.61	9.60	36.0	9.4	1.27	0.07
V	0.37	2.90	6.45	17.1	77.9	20.3	b.d.	–
VI	0.19	1.50	8.67	22.90	45.3	11.8	b.d.	–

[a] Values are tissue concentrations of peptides (pmol/mg protein) and percentage of the [Arg^8]vasopressin-(1–9) contents

radioimmunoassay system recognizing the C-terminal fragments of [Arg^8]vasopressin, a number of immunoreactive components were detected in the hippocampus, amygdala, septum, medulla oblongata, and hypothalamus. These immunoreactive components coeluted with synthetic replicates of the in vitro identified [Arg^8]vasopressin fragments or their reduced forms, and were not detected by an N-terminal [Arg^8]vasopressin antiserum (Fig. 11). By coextraction of radiolabeled [Arg^8]vasopressin, the possibility that components arose during extraction was excluded.

These data suggest that C-terminal [Arg^8]vasopressin fragments are endogenously present in the brain. Immunological and chromatographical characteristics are consistent with a fragment structure similar to that of fragments generated in vitro. However, chemical characterization is lacking thus far, and will be required for a conclusion about the identity of these components.

Comparison of quantities of putative [Arg^8]vasopressin fragments showed that they were most abundant in proportion to [Arg^8]vasopressin in extrahypothalamic brain areas (Table 4). In comparison to [Arg^8]vasopressin, their levels were notably lower in the hypothalamus, while in the pituitary gland they were virtually undetectable (Table 4).

The discovery of substances resembling C-terminal [Arg^8]vasopressin fragments in the brain strongly suggests that the aminopeptidase-type processing of vasopressin apparent from in vitro studies is operative in situ as well. The high proportion of the immunoreactive substances in extrahypothalamic brain may indicate that metabolism of [Arg^8]vasopressin is mostly associated with the areas having vasopressinergic terminals. This would correspond with the idea that conversion can take place after release during exposure to the extracellular environment (Burbach and De Wied 1981; Burbach and De Kloet 1982). The absence of immunoreactive substances in the pituitary gland may indicate that this tissue, unlike the brain, lacks the appropriate peptidase(s). Alternatively, fragments formed extracellularly may be rapidly removed from the pituitary. In view of the central activities of fragments and their inability to exert peripheral effects, an enzyme system exclusively active in the brain could provide tissue specificity for the production of active metabolites.

4 Biologic Activities of Metabolites of Oxytocin, Vasopressin, and Vasotocin

Structure-activity studies on the central effects of oxytocin and vasopressin have formed the basis of the concept that these peptides enclose shorter active sequences in their structure (De Wied 1977, 1980; De Wied and Bohus 1978; Burbach and De Wied 1981). These studies originally focussed on two parts of the molecules: the N-terminal "ring," composed of residues 1–6, and the C-terminal "tail," comprising residues 7–9 (for primary structures, see Fig. 1). In structure-activity studies, using behavioral paradigms as a bioassay, synthetic fragments having the ring or the tail were tested; sometimes nonapeptide analogues were employed. For reports and reviews of this work the reader is referred to the extensive literature (e.g., De Wied 1976, 1977, 1979, 1980; De Wied and Bohus 1978; Walter et al. 1975, 1978; Kovacs et al. 1982). As a result of these studies, it has been concluded that the central and peripheral activities of oxytocin and vasopressin are structurally dissociated, and the hypothesis of fragments with more selective and/or potent central activities, of which the nonapeptides serve as precursors, has been put forward (De Wied 1977, 1980; De Wied and Bohus 1978).

The recent identification of the main metabolites of oxytocin and vasopressin, generated in vitro by brain peptidases, prompted the further evaluation of these postulates by studies on their central and peripheral activities. Moreover, a physiological role for the metabolites and functional aspects of the proteolytic processing could be derived from such experiments. Current data are summarized in Sect. 4.1 and 4.2. For biologic experiments, several of the metabolites and their des-GlyNH_2 derivatives have been synthesized (Van Nispen et al. 1983, 1984). Peripheral and central activities have been tested.

4.1 Peripheral Activities

Oxytocin and vasopressin have well-known peripheral hormonal activities. In order to screen for peripheral effects of metabolites, bioassays for the classic endocrine effects have been employed.

Uterotonic activity of peptides was tested in vitro on the isolated rat uterus pretreated with estradiol benzoate (Overweg et al. 1968). The lowest effective dose of oxytocin in this test was 0.14 ng/ml. Doses of [pGlu^4,Cyt^6]oxytocin-(4-9) ranging from 0.001 to 100 ng/ml did not have uterotonic action (Burbach et al. 1983a, c). The uterotonic effect of oxytocin was not inhibited by a 1000-fold molar excess of the [pGlu^4]-(4-9) fragment. Vasopressin activity of [pGlu^4,Cyt^6,Arg^8]vasopressin-(4-9) and its 4-8 derivative was tested in the in vitro bioassay of Dekanski (1952). [Arg^8]vasopressin in a dose of 10 ng caused a significant rise in blood pressure. The two fragments were inactive in doses ranging from 10 to 500 ng (Burbach et al. 1983b, c). No antagonistic effect was observed. After subcutaneous injection of 3 ng [pGlu^4,Cyt^6,Arg^8]vasopressin-(4-8), no effect on blood pressure was noted (De Wied et al. 1984a, b).

The currently available data show that the [pGlu^4]-(4-9) fragments of oxytocin and [Arg^8]vasopressin are devoid of the classic endocrine activities of the nonapeptides. This lack of hormonal activity is in agreement with the structural requirements of oxytocin and vasopressin for hormonal activity as defined by Walter and colleagues (Walter 1977; Walter et al. 1977). The required structural elements include the intact N-terminal ring structure and residues 2, 3, and 5. Other effects of metabolites on possible peripheral targets have not yet been investigated in detail and cannot yet be ruled out. Preliminary observations indicate that the vasopressin fragments are not able to stimulate the release of adrenocorticotropic hormone (ACTH) from the pituitary gland, nor to potentiate the effect of corticotropin-release factor (CRF) (F. Antoni, unpublished). The inability of [pGlu^4,Cyt^6,Arg^8]vasopressin-(4-9), [Cyt^6,Arg^8]vasopressin-(3-9), and -(5-9) to displace ^{3}H-vasopressin from binding sites in anterior pituitary membranes of normal and Brattleboro rats in vitro, investigated by B. Koch (unpublished), supports this finding. The oxytocin fragments were inactive in stimulating the release of α-MSH from the perfused frog neurointermediate lobe (M. Lamacz, M.-C. Tonon and M. Vaudry, unpublished).

4.2 Central Activities

The activity of metabolites in affecting brain functions was investigated in a passive avoidance situation (Ader et al. 1972; Ader and De Wied 1972; see also legend

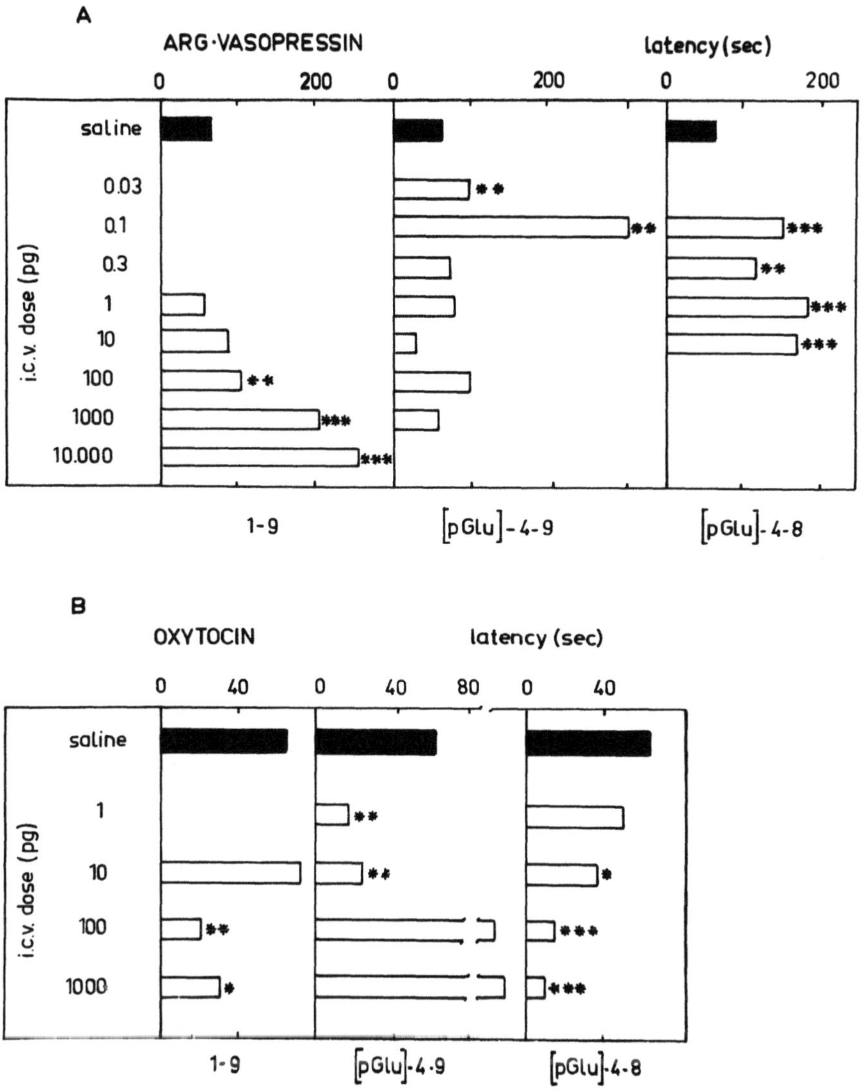

Fig. 12 A, B. Activity of the [pGlu^4]-(4-9) fragments of [Arg^8]vasopressin **A** and oxytocin **B** and [pGlu^4]-(4-8) derivatives on the consolidation of passive avoidance behavior. Passive avoidance behavior was studied in a one-trial learning, stepthrough situation according to Ader and De Wied (1972). Animals were placed on an illuminated platform attached to a dark compartment and were allowed to become habituated to the situation. At the end of the adaptation sessions, the animals received an inescapable electric shock to the foot (0.25 mA for 2 s) upon entering the dark compartment. Then peptides or saline were immediately administered by intracerebroventricular injection through a cannula placed in the right lateral ventricle. The latency with which the animals reenter the dark compartment after being placed once more on the platform 24 h later is a measure for the consolidation of passive avoidance behavior. The response of saline-treated animals is shown by the *black bars*. The *open bars* represent the latencies of peptide-treated animals.
* $p<0.05$; ** $p<0.01$; *** $p<0.001$; as compared to saline in the Mann-Whitney U-test

to Fig. 12), by intracerebroventricular administration of peptides in a wide dose-range (Fig. 12). The oxytocin metabolite [pGlu⁴,Cyt⁶]oxytocin-(4-9) has the same type of effect as oxytocin, i.e., attenuation of consolidation of passive avoidance behavior, but is about 100 times more potent (Burbach et al. 1983 a, c). The lowest effective dose of the peptide was 1 pg, while for oxytocin it was 100 pg. A U-shaped dose-response relationship was displayed by [pGlu⁴,Cyt⁶]oxytocin-(4-9). Its derivative [pGlu⁴,Cyt⁶]oxytocin-(4-8) had a more consistent effect in the tested doses, but was less potent (Fig. 12 B).

As does [Arg⁸]vasopressin, the metabolite [pGlu⁴,Cyt⁶,Arg⁸]vasopressin-(4-9) facilitates consolidation of passive avoidance behavior (Burbach et al. 1983 b, c). The peptide is approximately 1000-fold more potent than [Arg⁸]vasopressin after intracerebroventricular administration (Fig. 12 A). Doses of 0.03 and 0.1 pg of [pGlu⁴,Cyt⁶,Arg⁸]vasopressin-(4-9) were effective in facilitating the behavioral response of the animals. The lowest effective dose of [Arg⁸]vasopressin to elicit this effect was 100 pg. Again, a U-shaped dose-response relationship was observed with the [pGlu⁴]-(4-9) fragment. Its derivative [pGlu⁴,Cyt⁶,Arg⁸]vasopressin-(4-8) was active in all tested doses and was also centrally active after systemic administration in doses lower than those required for [Arg⁸]vasopressin (De Wied et al. 1984 a, b). In these studies the direct central action of the vasopressin fragment was demonstrated, and evidence for the existence of central vasopressin receptors with a specificity different from that of peripheral receptors was obtained (De Wied et al. 1984). In addition, [pGlu⁴,Cyt⁶,Arg⁸]vasopressin-(4-9) is more selective in its central activity than [Arg⁸]vasopressin. While central administration of [Arg⁸]vasopressin normalizes endotoxin-induced hyperthermia, the [pGlu⁴]-(4-9) fragment is inactive (Burbach et al. 1983 b, Kovács et al. 1983).

In recent structure-activity relationship studies with a series of C-terminal [Arg⁸]vasopressin fragments it appeared that central activities of fragments are not confined to [pGlu⁴,Cyt⁶,Arg⁸]vasopressin-(4-9) and -(4-8) (Gaffori et al. 1985). In particular [Cyt⁶,Arg⁸]vasopressin-(5-9) and -(5-8) are potently active in passive and active avoidance paradigms. [Cyt⁶,Arg⁸]vasopressin-(5-9) has been recognized as a [Arg⁸]vasopressin metabolite in in vitro and in vivo experiments (Wang et al. 1983a; Burbach et al. 1984a). After subcutaneous injection [Cyt⁶,Arg⁸]vasopressin-(5-9) and -(5-8) were the most potent peptides to facilitate consolidation and retrieval of a passive avoidance response. When given i.c.v. at least a 1000-fold lower dose of these peptides was sufficient to elicit similar effects on passive avoidance behavior (Gaffori et al. 1985). Comparison of the potencies of [Arg⁸]vasopressin-(1-9), [Cyt⁶,Arg⁸]vasopressin-(3-9), -(5-9), -(5-8), [pGlu⁴,Cyt⁶,Arg⁸]vasopressin-(4-9) and -(4-8) to affect consolidation and retrieval of passive avoidance behavior, extinction of active avoidance behavior and protection against pentylenetetrazol-induced retrograde amnesia indicated that the [Arg⁸]vasopressin metabolites differed in their selectivity to affect these behaviors (Gaffori et al. 1985).

Effect of [Arg⁸]vasopressin metabolites have been investigated in an electrophysiological study (Urban and Joëls 1985). It has been found that [Arg⁸]vasopressin-(1-9), topically applied to the septum, increased negativity of the field potential which was evoked in the lateral septum by fibria-fornix stimulation (Joëls and Urban 1984). Increase in negativity was also elicited by [pGlu^{4},Cyt^{6},Arg^{8}] vasopressin-(4-9), -(4-8) and [Cyt^{6},Arg^{8}]vasopressin-(5-9). However, at least 1000-fold higher concentrations of these metabolites were needed to elicit this effect.

The highly potent behavioral effects of C-terminal [Arg^{8}]vasopressin fragments suggests the existence of specific binding sites for these peptides in the brain. The possibility that the effects are directly mediated by binding to the putative central [Arg^{8}]vasopressin receptors is contradicted by two recent findings. Firstly, [pGlu^{4},Cyt^{6},Arg^{8}]vasopressin-(4-9) is much less effective to displace ^{3}H-vasopressin from binding sites in hippocampal membranes in vitro on brain slices than [Arg^{8}]vasopressin (Constantini and Pearlmutter 1984; De Kloet et al. 1985 b), suggesting that the fragment has low affinity to these [Arg^{8}]vasopressin receptors. Secondly, [pGlu^{4},^{35}S-Cyt^{6},Arg^{8}]vasopressin-(4-9) binds specifically to discrete brain regions in brain slices as shown by autoradiography (De Kloet et al. 1985 b). The distribution of ^{35}S-labeled binding sites is distinctly different from the distribution of putative receptors labeled with ^{3}H-vasopressin (Dorsa et al. 1983; Biegon et al. 1984; Brinton et al. 1984 a, b; De Kloet et al. 1985 a). Highest density of [pGlu^{4},^{35}S-Cyt^{6},Arg^{8}]vasopressin-(4-9) binding sites are localized in the pineal gland, organum vasculosum lamina terminalis (OVLT), arcuate nucleus and nucleus tractus solitarii. These findings suggest that receptors for C-terminal vasopressin fragments independent from those of vasopressin may exist in the brain (De Kloet et al. 1985 b).

It is of interest to note that the vasopressin metabolite [pGlu^{4},Cyt^{6},Arg^{8}]vasopressin-(4-9) is identical to the vasotocin metabolite [pGlu^{4},Cyt^{6}]vasotocin-(4-9) (Wang et al. 1983 a). This metabolite has a vasopressin-like effect on passive avoidance behavior in rats (facilitation), while vasotocin itself resembles oxytocin in this test (attenuation) (De Wied and Bohus 1978). Thus, proteolytic conversion of vasotocin leads to the generation of a metabolite with opposite central activity. This supports the hypothesis of De Wied and Bohus (1978), that in primitive organisms vasotocin may have both vasopressin- and oxytocin-like functions in the brain. Proteolysis can provide a mechanism to express these dual activities from a single peptide (Wang et al. 1983 a).

The data on the biologic activities of metabolites of oxytoxin and vasopressin thus far show that these peptides are centrally potent and selective in their central actions, but do not have apparent peripheral activities. This makes these metabolites candidates to represent neuropeptides in their own right.

5 Concluding Remarks

Proteolytic enzymes play a crucial role in peptidergic systems. Three functionally different roles can be distinguished.

Firstly, proteolytic enzymes mediate *precursor processing:* they cleave the inactive precursor forms and liberate peptides with biologic activities. Precursor processing is an intracellular process common to the biosynthesis of all known biologically active peptides.

Secondly, proteolytic enzymes mediate *metabolite conversion:* they act on peptides that are often active by themselves and generate shorter fragments with more potent or novel activities. Examples are found in the peripheral and central renin-angiotensin systems (Peach 1977; Ganten et al. 1983; Lang et al. 1983) and the central β-endorphin system (Burbach et al. 1980b; Burbach 1984).

Thirdly, proteolytic enzymes terminate biologic activities by the *inactivation and degradation* of peptides. The inactivation of enkephalins by brain peptidases is a well-elaborated case (Schwartz 1984).

The three types of proteolytic action have their own requirements with respect to compartmentalization and localization of proteolytic enzymes, peptide substrates, and bond specificity of the enzymes. Of the three types, precursor processing is as yet the best-defined process (Docherty and Steiner 1982; Lazure et al. 1983; Burbach 1984; Loh et al. 1984).

This review has been concerned with the cleavage of oxytocin, vasopressin, and related peptides by brain peptidases. The data give some indications in the direction of inactivating mechanisms, as well as metabolic conversions generating active fragments. Peptidases able to cleave oxytocin and vasopressin are found all over the brain and in most subcellular organelles. Due to the rigid molecular structure of the peptides, the N-terminal and the C-terminal tail are the two major sites of proteolytic attack. The distributions of N- and C-terminal-cleaving peptidases show regional and subcellular differences. In principle, both types of cleavage can lead to inactivation and degradation, in particular when proteolysis takes place at sites where products cannot escape or exert a biologic action (e.g., lysosomes). However, arguments in favour of functional metabolic conversions of oxytocin and vasopressin and the formation of active metabolites in the brain can be extracted from the reviewed data: (a) synaptic membranes convert oxytocin and vasopressin via a single major metabolic pathway, (b) products of this pathway are endogenously found in brain tissue, and (c) they have potent and selective central activities. Characteristics are:

1. Conversion is mediated by an aminopeptidase activity. This enzyme does not require disulfide reduction and generates peptides with an intact disulfide bridge.

2. Cyclization into pGlu occurs. As a consequence, the [pGlu^4]-(4-9) fragments of oxytocin and vasopressin are stable intermediates.

3. Endogenous forms of conversion products are present in the brain, but not in the pituitary gland. In the brain, concentrations markedly below the concentrations of the parent peptide are found. It may be assumed that such metabolites are transient products not able to accumulate to a large extent. Together with the high level of aminopeptidase activity in synaptosomal preparations, a synaptic or extracellular localization of the metabolic conversion may be suggested.

4. Fragments having the 4-9 sequence of oxytocin and vasopressin are 100- to 1000-fold more potent than their parents in their central action. Peripherally, they are inactive.

Points 3 and 4 suggest that metabolic conversion of oxytocin and vasopressin is limited to the brain, where it can modulate the activities of the oxytocin and vasopressin systems. In this respect, the central and peripheral systems may be differentiated. Conclusions about the functional metabolic conversion of oxytocin and vasopressin into active metabolites require further experimental work and novel approaches. Valuable tools could be found in antibodies against the aminopeptidases and active peptides, specific peptidase inhibitors, and protected peptide analogues. For development of these tools, detailed studies of the proteolytic conversion of oxytocin and vasopressin, the properties and regulation of the peptidases, and the biologic properties of metabolites are still demanded.

References

Abraham GN, Podell DN (1981) Pyrolutamic acid. Mol Cell Biochem 38:181–190

Acher R (1981) Evolution of neuropeptides. TINS 4:225–229

Ader R, De Wied D (1972) Effects of lysine vasopressin on passive avoidance learning. Psych Sci 29:46–48

Ader B, Weijnen JAWM, Moleman P (1972) Retention of a passive avoidance response as a function of the intensity and duration of electric shock. Psych Sci 26:125–128

Ang VTY, Jenkins JS (1984) Neurohypophysial hormones in the adrenal medulla. J Clin Endocrinol Metab 58:688–691

Bickel M, Kuhl H, Sioe Eng Tan J, Taubert HD (1972) Evidence of a sex-specific effect of testosterone and progesterone upon L-cystine aminopeptidase activity in the hypothalamus and paleopallium of the rat. Neuroendocrinology 9:321–331

Biegon A, Terlou M, Voorhuis ThD, De Kloet ER (1984) Arginine vasopressin bindings sites in rat brain: A quantitative autoradiographic study. Neurosci Lett 44:229–234

Boer GJ (1984) Vasopressin and brain development: Studies using the Brattleboro rat. Peptides (in press)

Boer GJ, Swaab DF, Uylings HBM, Boer K, Buijs RM, Velis DN (1980) Neuropeptides in rat brain development. Prog Brain Res 53:202–227

Bohus B (1977) Effect of desglycinamide-lysine vasopressin (DG-LVP) on sexually motivated T-maze behavior in the male rat. Horm Behav 8:52–61

Branda LA, Ferrier BM (1971) Degradation of oxytocin by human placental tissue. Am J Obstet Gynecol 109:943–947

Brinton RE, Wamsley JK, Gee KW, Wan YP, Yamamura HI (1984a) ^{3}H-Oxytocin binding sites in the rat brain demonstrated by quantitative light microscopic autoradiography. Eur J Pharmacol 102:365–367

Brinton RE, Gee KW, Wamsley JK, Davis TP, Yamamura HI (1984b) Regional distribution of putative vasopressin receptors in rat brain and pituitary by quantitative autoradiography. Proc Natl Acad Sci USA 81:7248–7252

Buijs RM (1978) Intra- and extrahypothalamic vasopressin and oxytocin pathways in the rat: pathways to the limbic system, medulla oblongata and spinal cord. Cell Tissue Res 192:423–435

Buijs RM, De Vries GJ, Van Leeuwen FW, Swaab DF (1983) Vasopressin and oxytocin distribution and putative functions in the brain. Prog Brain Res 60:115–122

Burbach JPH (1980) Proteolytic conversion of β-endorphin and oxytocin in rat brain. Thesis, University of Utrecht

Burbach JPH (1984) Action of proteolytic enzymes on lipotropins and endorphins: Biosynthesis, biotransformation, and fate. Pharmacol Ther 24:321–354

Burbach JPH, De Wied D (1981) Memory effects and brain proteolysis of neurohypophyseal hormones. In: Schlesinger DM (ed) Neurohypophyseal peptide hormones and other biologically active peptides. Elsevier/North-Holland Biomedical, Amsterdam, pp 69–87

Burbach JPH, De Kloet ER (1982) Proteolysis of β-endorphin in brain tissue. Peptides 3:451–453

Burbach JPH, Lebouille JLM (1983) Proteolytic conversion of arginine-vasopressin and oxytocin by brain synpatic membranes. Characterization of formed peptides and mechanism of proteolysis. J Biol Chem 258:1487–1494

Burbach JPH, De Kloet ER, De Wied D (1980a) Oxytocin biotransformation in the rat limbic brain: characterization of peptidase activities and significance in the formation of oxytocin fragments. Brain Res 202:401–414

Burbach JPH, Loeber JG, Verhoef J, Wiegant VM, De Kloet ER, De Wied D (1980b) Selective conversion of β-endorphin into peptides related to γ- and α-endorphin. Nature 282:96–97

Burbach JPH, Schotman P, De Kloet ER (1980c) Oxytocin biotransformation in the rat limbic brain: Chemical characterization of two oxytocin fragments and proposed pathway for oxytocin conversion. Biochem Biophys Res Commun 97:1005–1013

Burbach JPH, De Kloet ER, Schotman P, De Wied D (1981) Proteolytic conversion of β-endorphin by brain synaptic membranes: characterization of generated β-endorphin fragments and proposed metabolic pathway. J Biol Chem 256:12463–12469

Burbach JPH, Wang X-C, Van Ittersum M (1982) Difference in susceptibility of arginine-vasopressin and oxytocin to aminopeptidase activity in brain membranes. Biochem Biophys Res Commun 108:1165–1171

Burbach JPH, Bohus B, Kovács GL, Van Nispen JW, Greven HM, De Wied D (1983a) Oxytocin is a precursor of potent behaviourally active neuropeptide. Eur J Pharmacol 94:125–131

Burbach JPH, Kovács GL, De Wied D, Van Nispen JW, Greven HM (1983b) A major metabolite of arginine-vasopressin in the brain is a highly potent neuropeptide. Science 221:1310–1312

Burbach JPH, Kovács GL, Wang X-C, De Wied D (1983c) Metabolites of arginine-vasopressin and oxytocin are highly potent neuropeptides in the brain. In: Koch G, Richter D (eds) Biochemical and clinical aspects of neuropeptides: biosynthesis, processing and gene structure. Academic, New York, pp 211–224

Burbach JPH, Lebouille JLM, Wang X-C (1984a) Proteolytic conversion of neuropeptides into active fragments. In: Vizi ES, Magyar K (eds) Regulation of transmitter function: basic and clinical aspects. Akademiai Kiado, Budapest, pp 237–247

Burbach JPH, Wang X-C, Ten Haaf JA, De Wied D (1984b) Substances resembling C-terminal vasopressin fragments are present in the brain but not in the pituitary gland. Brain Res 306:384–387

Carraway R, Leeman SE (1975) The amino acid sequence of a hypothalamic peptide neurotensin. J Biol Chem 250:1907–1911

Celis ME (1975) Serum MSH levels and the hypothalamic enzymes involved in the formation of MSH-RF during the estrous cycle in the rat. Neuroendocrinology 18:256–262

Celis ME, Taleisnik S (1971a) Formation of a melanocyte-stimulating hormone release inhibiting factor by hypothalamic extracts from rats. Int J Neurosci 1:223–230

Celis ME, Taleisnik S (1971b) In vitro formation of a MSH-releasing agent by hypothalamic extracts. Experientia 27:1481–1482

Celis ME, Taleisnik S (1974) Estrogen influence on the hypothalamic enzymes involved in the formation of melanocyte-stimulating hormone release-inhibiting factor (MSH-R-IF). Proc Soc Exp Biol Med 145:142–144

Celis ME, Taleisnik S, Walter R (1971) Regulation of formation and proposed structure of the factor inhibiting the release of melanocyte-stimulating hormone. Proc Natl Acad Sci USA 68:1428–1433

Constantini MG, Pearlmutter AF (1984) Properties of the specific binding sites for arginine-vasopressin in rat hippocampal synaptic membranes. J Biol Chem 259:11739–11745

Cooper KE, Kasting NW, Lederis K, Veale WC (1979) Evidence supporting a role for vasopressin in natural suppression of fever in sheep. J Physiol (Lond) 295:33–45

Dekanski J (1952) The quantitative assay of vasopressin. Br J Pharmacol 7:567–572

De Kloet ER, Rotteveel F, Voorhuis ThAM, Terlou M (1985a) Topography of binding sites for neurohypophyseal hormones in rat brain. Eur J Pharmacol 110:113–119

De Kloet ER, Voorhuis ThAM, Burbach JPH, De Wied D (1985b) Autoradiographic localization of binding sites for the arginine-vasopressin metabolite AVP-(4-9) in rat brain. Neurosci Lett 56:7–11

Deslauriers R, Smith ICP (1970) Evidence from proton magnetic resonance data for the stacking of aromatic amino acids in lysine vasopressin: comparison with oxytocin derivatives and related peptides. Biochem Biophys Res Commun 40:179–185
De Wied D (1965) The influence of the posterior and intermediate lobe of the pituitary and pituitary peptides on the maintenance of a conditioned avoidance response in rats. Int J Neuropharmacol 4:157–167
De Wied D (1969) Effects of peptide hormones on behavior. In: Ganong WF, Martini L (eds) Frontiers in neuroendocrinology. Oxford University Press, Oxford, pp 97–140
De Wied D (1971) Long term effect of vasopressin on the maintenance of a conditioned avoidance response in rats. Nature 232:58–60
De Wied D (1976) Behavioral effects of intraventricularly administered vasopressin and vasopressin fragments. Life Sci 19:685–690
De Wied D (1977) Peptides and behavior. Life Sci 20:195–204
De Wied D (1979) Neurohypophyseal hormones and memory. In: Laszlo FA (ed) Recent results in peptide hormones and androgenic steroid research. Akadémiai Kiadó, Budapest, pp 149–154
De Wied D (1980) Behavioural actions of neurohypophyseal peptides. Proc R Soc Lond [Biol] 210:183–195
De Wied D (1983) Central actions of neurohypophyseal hormones. Prog Brain Res 60:155–167
De Wied D, Bohus B (1978) The modulation of memory processes by vasotocin, the evolutionary oldest neurosecretory principle. Prog Brain Res 48:327–334
De Wied D, Greven HM, Lande S, Witter A (1972) Dissociation of the behavioural and endocrine effects of lysine vasopressin by tryptic digestion. Br J Pharmacol 45:118–122
De Wied D, Gaffori O, Van Ree JM, De Jong W (1984a) Central target for the behavioural effects of vasopressin neuropeptides. Nature 308:276–278
De Wied D, Gaffori O, Van Ree JM, De Jong W (1984b) Vasopressin antagonists block peripheral as well as central vasopressin receptors. Pharmacol Biochem Behav 21:393–400
De Wildt D, Verhoef J, Witter A (1982) H-Pro-[^{3}H]Leu-Gly-NH_2: Uptake and metabolism in rat brain. J Neurochem 38:67–74
Docherty K, Steiner DF (1982) Post-translational proteolysis in polypeptide hormone biosynthesis. Ann Rev Physiol 44:624–638
Dorsa DM, Majumbar LA, Petracca FM, Baskin DG, Cornett LE (1983) Characterization and localization of ^{3}H-arginine vasopressin binding to rat kidney and brain tissue. Peptides 4:699–704
Ferrier BM, Branda LA (1966) Plasma oxytocinase. Proceedings of the 3rd international pharmacological congress, Sao Paulo, p 179
Feuerstein G, Zerbe RL, Faden AL (1984) Central cardiovascular effects of vasotocin, oxytocin and vasopressin in conscious rats. J Pharmacol Exp Ther 228:348–358
Folkers K, Cheng J-K, Curry L, Bowers CY, Weil A, Schally AV (1970) Synthesis and relationship of L-glutaminyl-L-histidyl-L-prolinamide to the thyrotropin releasing hormone. Biochem Biophys Res Commun 39:110–113
Frith DA, Hooper KC (1971a) The effect of coitus on the activity of certain enzymes in the female rabbit hypothalamus. Acta Endocrinol 66:213–220
Frith DA, Hooper KC (1971b) The action of some ovulation inhibitors on the rabbit hypothalamus. Acta Endocrinol 66:221–228
Frith DA, Hooper KC (1975) The development of an in vitro system for estimating oxytocinase in the rabbit hypothalamus and the effects of some protein synthesis inhibitors and oestradiol-17β on enzyme activity. Acta Endocrinol 75:443–451
Gaffori O, Burbach JPH, Kovács GL, Van Ree JM, De Wied D (1985) Structure activity relationship studies with C-terminal fragments of vasopressin and oxytocin on avoidance behavior. Submitted
Ganten D, Hermann K, Bayer C, Unger Th, Lang RE (1983) Angiotensin synthesis in the brain and increased turnover in hypertensive rats. Science 221:869–871

Grant NH, Clark DE, Rosanoff EI (1973) Evidence that Pro-Leu-Gly-NH_2, tocinoic acid and des-Cys-tocinoic acid do not affect secretion of melanocyte-stimulating hormone. Biochem Biophys Res Commun 51:100–106
Griffith EC, Hooper KC (1972a) The effect of ovariectomy on the activity of certain enzymes in the female rat hypothalamus. Acta Endocrinol 69:249–256
Griffith EC, Hooper KC (1972b) The effect of neonatal androgen on the activity of certain enzymes in the rat hypothalamus. Acta Endocrinol 70:767–774
Griffith EC, Hooper KC (1973a) The effects of orchidectomy and testosterone propionate injection on peptidase activity in the male rat hypothalamus. Acta Endocrinol 72:1–8
Griffith EC, Hooper KC (1973b) Peptidase activity in the hypothalami of rats treated neonatally with oestrogens. Acta Endocrinol 72:9–17
Griffith EC, Hooper KC (1973c) Changes in hypothalamic peptidase activity during the estrous cycle in the adult female rat. Acta Endocrinol 74:41–48
Griffith EC, Hooper KC (1973d) The stimulating influence of gonadal steroids and metalibine (ICI 33,828) on peptidase activity in the female rat hypothalamus. Fertil Steril 24:269–274
Griffith EC, Hooper KC (1974) Competitive inhibition between oxytocin and LH-RH for the same enzyme system in the rat hypothalamus. Acta Endocrinol 75:435–442
Griffith EC, Hooper KC, Hopkinson CRN (1973) Evidence for an enzymic component in the rat hypothalamus capable of inactivating luteinizing hormone releasing factor (LRF). Acta Endocrinol 74:49–55
Heil H, Meltzer V, Kuhl H, Abraham R, Taubert HD (1971) Stimulation of L-cystine-aminopeptidase activity by hormonal steroids and steroid-analogs in the hypothalamus and other tissues of the female rat. Fertil Steril 22:181–187
Heller H, Urban FF (1935) The fate of the antidiuretic principle of postpituitary extracts in vivo and in vitro. J Physiol (Lond) 85:502–518
Hoffman PL (1982) Structural requirements for neurohypophyseal peptide maintenance of ethanol tolerance. Pharmacol Biochem Behav 17:685–690
Hoffman PL, Walter R (1976) Preparation of oxytocin specifically ^{14}C-labeled in the tyrosine residue. FEBS Lett 66:176–178
Hooper KC (1962) The catabolism of some physiologically active polypeptides by homogenate of dog hypothalamus. Biochem J 83:511–517
Hooper KC (1963) The enzymic inactivation of some physiologically active polypeptides by different parts of the nervous system. Biochem J 88:399–404
Hooper KC (1964) The distribution of hypothalamic peptidases in pregnant and non-pregnant dogs. Biochem J 90:584–587
Hooper KC (1966a) Some observations on the behavior of hypothalamic enzymes during the time of blastocyst implantation in the rabbit. Biochem J 99:128–133
Hooper KC (1966b) The metabolism of oxytocin during lactation in the rabbit. Biochem J 100:823–826
Hooper KC (1968) The effects of ovariectomy and injected oestradiol monobenzoate on polypeptide metabolism in the hypothalamus. Biochem J 110:151–153
Hui KS, Cheng K-P, Wang K-H, Salshutz M, Lajtha A (1980) Degradation of melanotropin inhibiting factor by brain. J Neurochem 35:471–478
Ivell R, Richter D (1984) Structure and comparison of the oxytocin and vasopressin genes from rat. Proc Natl Acad Sci USA 81:2006–2010
Joëls M, Urban IJA (1984) Arginine^8-vasopressin enhances the responses of lateral septal neurons in the rat to excitatory amino acids and fimbriafornix stimuli. Brain Res 311:201–209
Kastin AJ, Laurence SP, Coy DH (1981a) Radioimmunoassay of MIF-1/Tyr-MIF-1-like material in rat pineal. Pharmacol Biochem Behav 13:901–905
Kastin AJ, Laurence SP, Coy DH (1981b) Radioimmunoassayable N-Tyr-MIF-1-like activity in rat brain is increased by pinealectomy. Brain Res Bull 7:697–702
Kasting NW, Veale WL, Cooper KE (1980) Convulsive and hypothermic effects of vasopressin in the brain of the rat. Can J Physiol Pharmacol 58:316–319
Koida M, Walter R (1976) Post-proline cleaving enzyme: purification of this endopeptidase by affinity chromatography. J Biol Chem 251:7593–7599

Koida M, Glass JD, Schwartz IL, Walter R (1971) Mechanism of inactivation of oxytocin by rat kidney enzymes. Endocrinology 88:633–643
Kovács GL, De Wied D (1983) Hormonally active arginine-vasopressin suppresses endotoxin-induced fever in rats: lack of effect of oxytocin and a behaviorally active vasopressin fragment. Neuroendocrinology 37:258–261
Kovács GL, Bohus B, Versteeg DHG, Telegdy G, De Wied D (1982) Neurohypophyseal hormones and memory. In: Yoshida H, Hagihara Y, Ebashi S (eds) Advances in pharmacology and therapeutics II, vol 1. Pergamon, Oxford/New York, pp 175–187
Kuhl H, Taubert HD (1975a) Inactivation of luteinizing hormone releasing hormone by rat hypothalamic L-cystine arylamidase. Acta Endocrinol 78:634–648
Kuhl H, Taubert HD (1975b) Short-loop feedback mechanism of luteinizing hormone: LH stimulates hypothalamic L-cystine arylamidase to inactivate LH-RH in the rat hypothalamus. Acta Endocrinol 78:649–663
Kuhl H, Rosniatowski C, Oen S-A, Taubert HD (1974a) Sex steroids stimulate the activity of hypothalamic arylamidases in the rat. Acta Endocrinol 76:1–14
Kuhl H, Rosniatowski C, Bickel M, Taubert HD (1974b) Stimulation by steroids and age-dependent changes of L-cystine arylamidase activity in the hypothalamus of the rat. Acta Endocrinol 76:15–23
Kuhl H, Rosniatowski C, Taubert HD (1977) The regulatory function of a pituitary LH-RH degrading enzyme system in the feedback control of gonadotrophins. Acta Endocrinol 86:60–70
Kuhl H, Rosniatowski C, Taubert HD (1978) The activity of an LH-RH degrading enzyme in the anterior pituitary during the rat oestrus cycle and its alteration by injections of sex hormones. Acta Endocrinol 87:476–484
Land H, Schütz G, Schmale H, Richter D (1982) Nucleotide sequence of cloned cDNA encoding bovine arginine vasopressin-neurophysin II precursor. Nature 295:299–303
Land H, Grez M, Ruppert S, Schmale H, Rehbein M, Richter D, Schütz G (1983) Deduced amino acid sequence from the bovine oxytocin-neurophysin I precursor cDNA. Nature 302:342–344
Lang RE, Gaida W, Ganten D, Hermann K, Kraft K, Unger Th (1983) Neuropeptides and central blood pressure regulation. In: Ganten D, Pfaff D (eds) Central cardiovascular control. Springer, Berlin Heidelberg New York, pp 103–124 (Current topics in neuroendocrinology, vol 3)
Lauson HD (1974) Metabolism of the neurohypophyseal hormones. In: Greep RO, Astwood EB, Kobil E, Sawyer WH (eds) The pituitary gland and its neuroendocrine control. American Physiological Society, Washington, pp 287–293 (Handbook of physiology, vol 4, part 1)
Lazure C, Seidah NG, Pélaprat D, Chrétien M (1983) Proteases and post-translational processing of prohormones: a review. Can J Biochem 61:501–515
Lim AT, Lolait SJ, Barlow JW, Autelitano DJ, Toh BH, Boublik J, Abraham J, Johnston CJ, Funder JW (1984) Immunoreactive arginine-vasopressin in Brattleboro rat ovary. Nature 310:61–64
Loh YP, Brownstein MJ, Gainer H (1984) Proteolysis in neuropeptide processing and other neural functions. Ann Rev Neurosci 7:189–222
Manberg PL, Youngblood WW, Kizer JS (1982) Developments of a radioimmunoassay for Pro-Leu-Gly-NH_2 (PLG or HIF-1): Evidence that PLG is not present in rat brain. Brain Res 241:279–284
Marks N, Walter R (1972) MSH-release inhibiting factor: Inactivation by proteolytic enzymes. Proc Soc Exp Biol Med 140:673–676
Marks N, Abrash L, Walter R (1973) Degradation of neurohypophyseal hormones by brain and purified brain enzymes. Proc Soc Exp Biol Med 142:455–460
Marks N, Galoyan A, Grynbaum A, Lajtha A (1974) Protein and peptide hydrolases of the rat hypothalamus and pituitary. J Neurochem 22:735–739
Meisenberg G, Simmons WH (1984) Factors involved in the inactivation of vasopressin after intracerebroventricular injection in mice. Life Sci 34:1231–1240
Nair RMG, Kastin AJ, Schally AV (1971) Isolation and structure of hypothalamic MSH release inhibiting hormone. Biochem Biophys Res Commun 43:1376–1381

Nicholson HD, Swann RW, Burford GD, Wathes DC, Porter DG, Pickering BT (1984) Identification of oxytocin and vasopressin in the testis and in adrenal tissue. Regul Pept 8:141–146
Overweg NJA, Schwartz JL, Dubois BM, Walter R (1968) Inhibition of the action of oxytocin on the rat uterus by acyclic oxytocin intermediates. J Pharmacol Exp Ther 161:342–347
Peach MJ (1977) Renin-angiotensin system: biochemistry and mechanism of action. Physiol Rev 57:313–370
Pedersen CA, Prange AJ (1979) Induction of maternal behavior in virgin rats after i.c.v. administration of oxytocin. Proc Natl Acad Sci USA 76:6661–6665
Pedersen CA, Ascher JA, Monroe YL, Prange A Jr (1982) Oxytocin induces maternal behavior in virgin female rats. Science 216:648–649
Pliška V, Barth T, Thorn NA (1971 a) Some metabolites of specifically tritium-labeled lysine-vasopressin: identification by thin layer chromatography. Acta Endocrinol 67:1–11
Pliška V, Thorn NA, Vilhardt H (1971 b) In vitro uptake and breakdown of tritiated lysine-vasopressin by bovine neurohypophyseal and corticol tissue. Acta Endocrinol 67:12–22
Richter D (1983) Vasopressin and oxytocin are expressed as polyproteins. Trends Biochem Sci 8:278–281
Ruppert S, Scherer G, Schütz G (1984) Recent gene conversion involving bovine vasopressin and oxytocin precursor genes suggested by nucleotide sequence. Nature 308:554–557
Sawyer WH (1964) Vertebrate neurohypophyseal principles. Endocrinology 75:981–990
Sawyer WH (1977) Evolution of neurohypophyseal hormones and their receptors. Fed Proc 36:1842–1847
Schlank H, Walter R (1972) Enzymatic cleavage of post-proline peptide bonds: degradation of arginine-vasopressin and angiotensin II. Proc Soc Exp Biol Med 141:452–455
Schmale H, Richter D (1984) A single base deletion in the vasopressin gene is the cause of the diabetes insipidus in Brattleboro rats. Nature 308:705–709
Schmale H, Heinsohn S, Richter D (1983) Structural organization of the rat gene for the arginine vasopressin-neurophysin precursor. EMBO J 2:763–767
Schwartz JCH, Gras C, Giras B, Llorens C, Malfroy B, Rose Ch, Zurel K (1984) Metabolic inactivation of enkephalins. In: Vizi ES, Magyar K (eds) Regulation of transmitter function. Akadémiai Kiadó, Budapest, pp 217–227
Simmons WH, Brecher AS (1973) Inactivation of melanocyte-stimulating hormone-release-inhibiting factor by a manganese-stimulated bovine brain aminopeptidase. J Biol Chem 248:5780–5784
Simmons WH, Walter R (1980) Carboxamidopeptidase purification and characterization of a neurohypophyseal hormone inactivating peptidase from toad skin. Biochemistry 19:39–48
Simmons WH, Walter R (1981) Enzyme inactivation of oxytocin: properties of carboxamidopeptidases. In: Schlesinger DH (ed) Neurohypophyseal peptide hormones and other biological active peptides. Elsevier North Holland, Amsterdam, pp 151–165
Sjöholm I, Yman L (1967) Degradation of oxytocin, lysine-vasopressin, angiotensin II and angiotensin-II-amide by oxytocinase (cystine aminopeptidase). Acta Pharm Suec 4:65–76
Smith CW (1981) Conformation-activity studies on oxytocin and vasopressin: exploring the roles of the moieties within the hydrophilic cluster. In: Schlesinger DH (ed) Neurohypophyseal peptide hormones and other biologically active peptides. Elsevier North Holland, Amsterdam, pp 23–36
Södersten P, Henning M, Melin P, Ludin S (1983) Vasopressin alters female sexual behaviour by acting on the brain independently of alterations in blood pressure. Nature 301:608–610
Sofroniew MV (1980) Projections from vasopressin, oxytocin and neurophysin neurons to neural targets in the rat and human. J Histochem Cytochem 28:475–478

Sofroniew MV, Weindl A (1978) Projections from the parvocellular vasopressin and neurophysin containing neurons of the suprachiasmatic nucleus. Am J Anat 153:391–430
Stahl GL, Walter R (1978) Synthesis of (2-C-14(U)tyrosine-arginine vasotocin with high specific radio and biological activities. J Label Comp Radiopharm 14:881–886
Taubert HD, Heil H, Kuhl H (1970) Untersuchungen über die Wirkung von Äthinylöstradiol auf die L-Cystin-Aminopeptidase Aktivität im Hypothalamus der weiblichen Ratte. In: Kracht J (ed) Endokrinologie der Entwicklung und Reifung. Proceedings of the 16th Symposium of the Deutsche Gesellschaft für Endokrinologie. Springer, Berlin Heidelberg New York, pp 288–289
Thody AJ, Lever de Vries CH, Tilders FJH (1980) The failure of L-prolyl-L-leucyl-glycinamide to inhibit the release of α-melanocyte stimulating hormone in the rat. Acta Endocrinol 93:300–305
Tonon MC, Leboulanger F, Delarne C, Jegou S, Fresel J, Leroux P, Vaudry H (1979) TRH as MSH-releasing factor in the frog. In: McKern A, Julisz M (eds) Biomedical endocrinology: synthesis and release of adenohypophyseal hormones, cellular and molecular mechanisms. Plenum, New York, pp 731–751
Tuppy H, Wiesbauer U, Wintersberger E (1962) Aminosäure-p-nitroanilide als Substrate für Aminopeptidasen und andere proteolytische Fermente. Hoppe Seylers Z Physiol Chem 329:278–288
Urban IJA, Joëls M (1985) Studies on function of arginine^{8}-vasopressin in lateral septum of rats. Proc XVIth Int Congress ISPNE, Kyoto, April 14–18, 1985, abstract FC-3-10
Van Nispen JW, Hannink JAJ, Greven HM (1983) Synthesis and behavioral activity of fragments of oxytocin and arginine vasopressin containing a cystine residue in position 6. In: Hruby VJ, Rich DH (eds) Peptides: structure and function. Proceedings of the 6th American peptide symposium. Pierce Chemical Company, pp 421–424
Van Nispen JW, Hannink JAJ, Schoffelmeer MS, Janssen WPA, Polderdijk JP, Greven HM (1984) Synthesis of fragments of arginine vasopressin and oxytocin containing a cystine residue in position 6. Recl Trav Chim Pays Bas 103:68–74
Van Ree JM (1982) Neurohypophyseal hormones and addiction. In: Hoshida H, Hagihara Y, Ebashi S (eds) Advances in pharmacology and therapeutics II, vol 1. CNS pharmacology – neuropeptides. Pergamon, Oxford, pp 199–209
Van Ree JM, De Wied D (1981) Vasopressin, oxytocin and dependence on opiates. In: Martinez JL et al. (eds) Endogenous peptides and learning and memory processes. Academic, New York, pp 397–412
Veale WC, Kasting NW, Cooper KE (1981) Arginine vasopressin and endogenous antipyresis: evidence and significance. Fed Proc 40:2750–2753
Versteeg CAM, Cransberg K, De Jong W, Bohus B (1983) Reduction of centrally induced pressor response by neurohypophyseal peptides: the involvement of lower brainstem mechanisms. Eur J Pharmacol 94:133–140
Walter R (1972) The role of enzymes in the formation and inactivation of peptide hormones. In: Hanson H, Jakubke HD (eds) Peptides. Elsevier/North Holland, Amsterdam, pp 363–378
Walter R (1974) Oxytocin and other peptide hormones as prohormones. In: Hatotani N (ed) Psychoneuroendocrinology. Karger, Basel, pp 285–294
Walter R (1977) Identification of sites in oxytocin involved in uterine receptor recognition and activation. Fed Proc 36:1872–1878
Walter R, Havran RT (1971) Arginine-vasopressin, lysine-vasopressin and oxytocin ^{14}C-labeled in the glycine residue. Experientia 27:645–646
Walter R, Schlank H (1975) Differences in the enzymatic inactivation of arginine-vasopressin and oxytocin by rat kidney homogenate. Endocrinology 96:811–814
Walter R, Simmons WH (1977) Metabolism of neurohypophyseal hormones: consideration from a molecular viewpoint. In: Moses AM, Share L (eds) Neurohypophysis. Karger, Basel, pp 167–188
Walter R, Griffith EC, Hooper KC (1973) Production of MSH-release-inhibiting hormone by a particulate preparation of hypothalami; mechanisms of oxytocin inactivation. Brain Res 60:449–457

Walter R, Hoffman PL, Flexner JB, Flexner LB (1975) Neurohypophyseal hormones, analogs and fragments: their effect on puromycin-induced amnesia. Proc Natl Acad Sci USA 72:4180–4184

Walter R, Smith C, Mehta PK, Boonjarern S, Arruda JAL, Kurtzmann NA (1977) Conformational considerations of vasopressin as a guide to the development of biological probes and therapeutic agents. In: Adreoli TE, Ganthar JJ, Recor FC (eds) Disturbances in body fluid osmolality. American Physiological Society, Bethesda, pp 1–36

Walter R, Van Ree JM, De Wied D (1978) Modification of conditioned behavior of rats by neurohypophyseal hormones analogues. Proc Natl Acad Sci USA 75:2493–2496

Wang X-C, Burbach JPH (1986) Formation of metabolites of [Arg^{8}]vasopressin (AVP) by brain peptidases: conversion of the intermediate [Cyt^{6}]AVP-(3-9). FEBS Lett. In press

Wang X-C, Burbach JPH, Verhoef J, De Wied D (1983 a) Proteolytic conversion of arginine-vasotocin by synaptic membranes from rat and chicken brain. Brain Res 275:83–90

Wang X-C, Burbach JPH, Verhoef J (1983 b) Proteolysis of adrenocorticotropin in brain: characterization of cleavage sites by peptidases in synaptic membranes and formation of peptide fragments. J Biol Chem 258:7942–7947

Wathes DC, Swann RW (1982) Is oxytocin an ovarian hormone? Nature 297:225–227

Zimmermann EA, Nilaver G, Hou-You A, Silverman AJ (1984) Vasopressinergic and oxytocinergic pathways in the central nervous system. Fed Proc 43:91–96

Oxytocin and Behavior

G. L. Kovács

Contents

1 Introduction

The past 10 years have witnessed the growth of the view that neuropeptides are essential to the functional integrity of the central nervous system. Oxytocin (OXT), which was long considered to be implicated in milk ejection only, is a neuropeptide with potent behavioral effects. Since the original discovery by

Institute of Pathophysiology, University Medical School, Semmelweis u.l., POB 531, Szeged, Hungary (H-6701)

Current Topics in Neuroendocrinology, Vol. 6

Sterba (1974), a great number of morphological (for reviews, see Buijs 1983; Swanson and Sawchenko 1983; Sofroniew 1983; Palkovits and Brownstein 1983; Kozlowski et al. 1983) and biochemical (Dogterom et al. 1978; Mens et al. 1983; Kovács et al. 1985d; Hawthorn et al. 1984) results indicate that the biologically active OXT is present in various extrahypothalamic (mainly limbic and brainstem) brain regions. The release of extrahypothalamic OXT by depolarizing stimuli has been demonstrated (Buijs and Van Heerikhuize 1982), and the existence of specific binding sites for OXT (putative OXT receptors) has been described in limbic brain structures (Ferrier et al. 1983; Brinton et al. 1984). However, the biological significance of OXT in the brain is not clear. The neuropeptide has been implicated in the regulation of behavioral reactions. Evidence has accumulated that OXT attenuates learning and memory processes (for review, see Kovács and Telegdy 1982), regulates the adaptive response to narcotic analgesics (Kovács et al. 1984b, c), and alters the efficacy of addictive drugs (Van Ree and De Wied 1977a; Kovács and Telegdy 1984; Kovács et al. 1985a). Central nervous OXT has also been implicated in the onset of maternal behavior (Pedersen and Prange 1979). In high doses, central OXT treatment causes behavioral arousal characterized by a variety of – partly toxic – symptoms (Kruse et al. 1977; Meisenberg 1981). Some of the above-mentioned behavioral effects of OXT are consequences of interactions of the neuropeptide with dopaminergic, noradrenergic, and serotoninergic transmitter pathways in the brain (for reviews, see Telegdy and Kovács 1979a, b; Kovács and Telegdy 1982; Versteeg 1983), which indicates that OXT may act as a neuromodulator (regulator) in the central nervous system. The present chapter is devoted to a review of the role of OXT in behavioral reactions and to an analysis of the possible involvement of catecholaminergic and serotoninergic neurotransmission in mediating behavioral effects of OXT.

2 Oxytocin in Learning and Memory Processes

2.1 Effects of Oxytocin on Fear-Motivated Active and Passive Avoidance Behavior

Research on the role of OXT in behavioral processes was particularly motivated by the pioneering data of De Wied and his associates indicating disturbances of learning in rats after removal of the posterior lobe of the pituitary (De Wied 1965). Subsequent studies revealed that the disturbed behavior of posterior lobectomized animals could be normalized with vasopressin (De Wied and Bohus 1966; De Wied 1969), and that vasopressin facilitates memory processes in intact animals too (De Wied 1976; Van Wimersma Greidanus et al. 1975, 1983; Kovács et al. 1977, 1982a; Telegdy and Kovács 1979a, b; Bohus et al. 1972, 1978a, b; Schulz et al. 1974, 1976; Koob et al. 1981; Walter et al. 1978 etc.). The first publications concerning the effects of OXT on fear-motivated conditioned avoidance behavior were rather controversial. While lysine vasopressin (LVP) eliminated behavioral abnormalities of posterior lobectomized rats, subcutaneous injection of OXT failed to normalize behavior (De Wied and Bohus 1966). OXT was also ineffective in preventing retrograde amnesia, whereas vasopressin was found to be effective. Thuse OXT appeared to be inactive with respect to behavioral processes. In high doses, OXT tended to mimic the effect of vasopressin on the maintenance of conditioned avoidance behavior (De Wied and Gispen 1977).

In contrast to the previous findings, Schulz et al. (1974) reported that daily peripheral (intraperitoneal) injection of rats with a low dose of OXT (ca. 2×10^{-11} mol/animal) significantly facilitated the extinction rate of a bench-jumping avoidance reaction in rats. This effect of OXT was more pronounced in water-deprived rats, which have a high level of circulating vasopressin (Schulz et al. 1976). As a consequence, rats treated with OXT stopped responding to the conditioned signal earlier than control animals. Since the extinction of this avoidance reaction was delayed by vasopressin treatment, Schulz et al. (1974, 1976) concluded that OXT affects conditioned avoidance reactions in the opposite direction to vasopressin. Kovács et al. (1978) published similar findings on the retention of passive avoidance behavior learned at one trial. In this study rats were trained to avoid electric shocks to the feet by jumping onto a safe bench and remaining there. Twenty-four hours after a single learning session, each animal was placed on the bench and the latency to step down was measured. Synthetic OXT markedly attenuated the retention of the passive avoidance response, which, in contrast, was facilitated by LVP (Kovács et al. 1978). Not only peripheral OXT treatment (Schulz et al. 1974, 1976; Kovács et al. 1977) but also the intracerebroventricular injection of OXT (Bohus et al. 1978 b) has been observed to facilitate the extinction of the avoidance response and to attenuate the passive avoidance response in a step-through-type experimental paradigm. If OXT was injected directly into the cerebral ventricles, however, considerably lower doses were required (10^{-13} mol) in order to attenuate conditioned avoidance behavior (Bohus et al. 1978 a, b; Kovács et al. 1982 a; Kovács and De Wied 1983).

These experiments indicated that OXT is capable of modifying brain mechanisms involved in the regulation of conditioned avoidance behavior, and suggested that under appropriate experimental conditions OXT may exert behavioral effects opposite to those induced by vasopressin.

2.2 Temporal Aspects Indicate the Involvement of Memory Processes

Active and passive avoidance behavior are rather complex reactions in which learning and memory processes play outstanding roles. Thus it was of interest to study whether the disruptive effect of OXT on fear-motivated conditioned avoidance reactions is related to effects of the neuropeptide on learning and memory. Conventionally, the term "learning" refers to the biological responses of the organism in the acquisition of information, including the perception of sensory input and a great number of performance factors which operate during any (external or internal) environmental stimuli leading to memory formation.

Although the effects of OXT on learning have not been analyzed in all details, there is little evidence to suggest that learning per se is attenuated by OXT (Kovács et al. 1978; Telegdy and Kovács 1979 a, b). In this context those experiments are the most conclusive in which OXT treatment preceded behavioral testing. Neither the early period (Schulz et al. 1974, 1976) of learning in a bench-jumping avoidance stituation nor later phases of learning of a partially acquired response in an active avoidance pole-jumping paradigm (Bigl et al. 1977) were affected by the peripheral injection of OXT. The acquisition of a passive avoidance response learned at one trial was also unaffected by a dose of OXT which proved effective with respect to the retention of the same response in the same animals (Kovács et al. 1978). The only positive findings indicating the interference of OXT treatment with learning are those of Bohus et al. (1978 b), which showed that OXT slightly decreased the performance of rats during the later phases of acquisition.

Oxytocin is a neurohypophyseal hormone which is normally released under stressful circumstances (Mens et al. 1982; Gibbs 1984). Since fear-motivated avoidance behavior can be regarded as a chain of stressful events (Mens et al. 1982), the aforementioned studies can be criticized because they ignored the release of endogenous OXT in the nontreated control animals. This might explain why the acquisition phase of avoidance learning in general and the retention of certain behavioral reactions in particular (Sahgal and Wright 1984) were unaffected by OXT. Since there is an increasing body of (often controversial) evidence that vasopressin can affect the ability of naive rats to learn how to find food in a Y-maze (Sara et al. 1982) or how to find water (Ettenberg et al. 1983), these data argue for the necessity of experiments in which the influence of OXT on learning processes is tested under stress-free experimental conditions. The data so far, however, do not support the idea that learning per se is influenced by OXT.

Conventionally, the term "memory" refers to the biological responses which start and take place after the specific learning situation has terminated. McGaugh (1966, 1973) has presented a simple and experimentally testable model for memory processes in rodents. It is assumed that memory processes involve consolidation, storage, and retrieval processes and that the later recall of learned behavior can be modulated by endogenous and exogenous substances by interaction with either consolidation or retrieval. One-trial-learning experimental paradigms are useful test models for investigation of the effects of substances on memory consolidation and retrieval respectively. In general, a single learning trial is presented and the retention of passive avoidance is tested 24 h after the learning trial. Operationally, a treatment given shortly after the learning trial influences consolidation processes. Since consolidation is a time-dependent process which is active for a few hours only, it is necessary to demonstrate that a treatment which effectively modifies behavior when given immediately or shortly after the learning trial is ineffective if applied after this sensitive period has passed. In a retrieval experiment, in order to determine whether a compound can improve performance by enhancing the use of information already learned and stored, the compound is given shortly before the retention test (i.e., before the reexposure to the behavioral situation). Although the conclusions drawn on memory from such time-dependent studies has recently been questioned by some authors (Gash and Thomas 1983), this technique proved useful in analyses of the effects of exogenous drugs (McGaugh and Dawson 1971), endogenous opioid peptides (Kovács and De Wied 1981; Izquierdo et al. 1981), and vasopressin (De Wied et al. 1976, 1984) on memory processes.

Time-dependent changes have been described in the performance of passive avoidance behavior following the intracerebroventricular administration of OXT (Bohus et al. 1978a). An immediate postlearning injection of OXT resulted in the significant attenuation of passive avoidance behavior tested 24 h after the learning trial. When treatment with the neuropeptide was postponed until 3 h after the learning trial, the attenuation of passive avoidance was still observed in the 24-h retention test, but this effect was less pronounced than that of an injection given immediately after the learning trial. If treatment was given 6 h after the learning trial, however, OXT was ineffective. A second effectiveness peak appeared if the peptide was injected shortly before the retention trial. Thus the effect of OXT on passive avoidance behavior was clearly dependent on the interval between learning and peptide treatment and Bohus et al. (1978a) therefore interpreted these data as indicating that OXT attenuates memory consolidation and retrieval processes.

Although the fact that OXT and related peptides attenuate active and passive avoidance behavior has been confirmed by various authors (Schulz et al. 1974, 1976; Kovács et al. 1978, 1982a; Kovács and De Wied 1983; Telegdy and Kovács 1979a, b; Bohus et al. 1978a, b; Drago et al. 1981; etc.), the interpretation of these data is hampered by the complexity of the physiological and behavioral variables that may contribute to the attentuation of performance in an aversively motivated task. The first of these variables is pain perception and pain reactivity. Kovács et al. (1985b) tested the effects of wide dose ranges of OXT on the pain reactivity of the mouse, as assessed by the tail withdrawal response in a tail flick test, and found no evidence of an analgesic effect of OXT. In rats, however, Kordower and Bodnar (1984) reported that high doses of intracerebroventricularly injected OXT did appear to induce analgesia, but they concluded that this was related more to the toxic side effects (barrel rotation) during the testing period than to a profound action of OXT on pain perception itself. There is therefore very little evidence at the present time that OXT interferes with pain perception or pain reactivity. However, even if the neuropeptide were to do so, such an effect would not play any major role in behavioral paradigms in which the peptide is injected after the painful learning trial (e.g., one-trial learning passive avoidance reaction). Another possibility that has recently been discussed in relation to behavioral effects of vasopressin is whether or not this neuropeptide affects motivation and arousal (Sahgal and Wright 1984; Sahgal et al. 1982; Gash and Thomas 1983). Such changes could be responsible for the effects of OXT on fear-motivated behavioral reactions. Practically no data are available concerning the influence of OXT and related peptides on motivation and arousal. As far as autonomic correlates of motivation and arousal are concerned, the peripheral injection of OXT did not influence the bradycardiac response of rats during orientation (Hagan and Bohus 1984). Central (intracerebroventricular) OXT treatment, on the other hand, was found to affect the bradycardiac response to conditioned aversive stimuli (Bohus, personal communication), which response is thought to reflect the attentional demands of the situation (Obrist et al. 1970). According to Feuerstein et al. (1984), however, a wide range of doses of OXT had no effect on the heart rate or blood pressure in intact rats not exposed to behavioral tasks.

The effect of OXT on locomotor activity is a further variable that could interfere with the effect of the neuropeptide on avoidance reactions. Kovács et al. (1985 b) studied the influence of OXT in wide dose ranges on the locomotor activity of mice in the home environment and on the open-field activity in a novel environment, but found no significant effect. In rats, low doses of OXT were likewise ineffective with respect to the open-field activity (Schulz et al. 1974), and high doses of subcutaneously administered OXT did not affect any open-field parameters either (Xiao et al. 1984). Andrews et al. (1983), on the other hand, reported that high doses of OXT caused a transient decrease in the motor activity of rats. However, such a transient decrease cannot account for a long-lasting effect of OXT on fear-motivated avoidance behavior.

The failure of the neuropeptide to alter pain perception, motor activity, and the autonomic aspects of motivation/arousal indirectly supports the idea that the changes observed in fear-motivated avoidance paradigms are in fact related to the effect of OXT on the processing of information. Additional evidence in support of this notion is provided by electrophysiological observations. Landfield et al. (1972) have presented experimental evidence that the rhythmic slow (6–10 cps) activity of the hippocampus in the postlearning period may be a correlate of a brain state which is optimum for memory consolidation. Drugs which facilitate memory processes also increase the occurrence of theta activity in the postlearning period. Conversely, amnesic treatment decreases the incidence of theta activity (Landfield and McGaugh 1972). Bohus et al. (1978 b) and Urban (1981) found that OXT caused a shift toward lower frequencies in the power spectrum of hippocampal theta activity, suggesting that the activity of neurons generating the theta activity was decreased by OXT.

In conclusion, the temporal aspects of the effect of OXT on fear-motivated avoidance reactions, together with the effects of this neuropeptide on the hippocampal electrical activity, suggest that OXT attenuates memory consolidation and retrieval processes in laboratory animals.

2.3 Oxytocin and Retrograde Amnesia

A number of procedures disrupting the electrical and biochemical activity of the central nervous system after a training episode have been shown to produce retrograde amnesia, i.e., the performance of the animals in the retention session is largely impaired and thus the situation is similar to that when no training has taken place. The disruptive effect of various amnesic treatments is most probably related to the fact that a long-lasting trace of the experience is not completely fixed, consolidated, or coded at the time of the amnesic treatment. The most frequently used amnesic treatments are electroconvulsive shock (ECS) treatment, the administration of convulsive drugs (e.g., pentylenetetrazol), and the inhibition of cerebral protein synthesis (e.g., with anisomycin, puromycin, or cycloheximide) (McGaugh 1966; Bohus et al. 1982; Kovács et al. 1983 a; Judge and Quartermain 1982; etc.). Recovery from retrograde amnesia may be spontaneous, or it may be facilitated by behavioral (Quartermain et al. 1977), pharmacological (McGaugh et al. 1979), or hormonal (Rigter et al. 1974; Rigter and Van Riezen

1978; Bohus et al. 1982; Kovács et al. 1983a) manipulations. It has been reported (Kovács and Telegdy 1984; Kovács et al. 1985c) that post learning injection of OXT disrupts the passive avoidance response in rats to a similar extent to that following ECS treatment or protein synthesis inhibition. In agreement with the proposed effect of OXT as an amnesic (amnesia-inducing) neuropeptide, this peptide and its structurally related analogs were found to be ineffective in preventing or reversing amnesia induced by other amnesic treatments (for reviews, see Rigter and Van Riezen 1978; Kovács and Telegdy 1982).

2.4 Structure-Activity Studies of the Effects of Fragments and Analogs of Oxytocin on Avoidance Behavior

It has been recognized that the effects of OXT on fear-motivated avoidance behavior are not restricted to the hormonally active nonapeptide, and that fragments of OXT also exert potent behavioral effects (Table 1) (De Wied 1979; Kovács et al. 1982a). Interestingly, both the six-membered ring structure (tocinoic acid and its amide) and the C-terminal tripeptide of OXT (prolyl-leucyl-glycinamide, PLG, MIF-I) facilitate avoidance reactions in rats, and thus the concept has been put forward by De Wied (1979) that OXT may possess more than one behaviorally active core able to influence learning and memory processes. It is worthwhile to stress that the C-terminal tripeptide of OXT results in opposite changes in avoidance reactions to those induced by the intact neuropeptide. Moreover, the tripeptide also antagonizes retrograde amnesia induced by OXT (Kovács et al. 1985c) and by other types of amnesic treatment (Walter et al. 1975; Rigter and Van Riezen 1978; Kovács et al. 1983a).

Whereas the above-mentioned fragments of OXT improve the performance of the animals in avoidance situations, des-glycinamide-9-oxytocin, which is a hormonally inert fragment of the neuropeptide containing the N-terminal octapeptide, is as potent as the parent nonapeptide in attenuating a passive avoidance response (Drago et al. 1981; Kovács et al. 1982a). There is little evidence, however, that des-glycinamide-9-OXT is a peptide naturally occurring in the brain. Efforts have therefore been made to investigate the brain metabolism of OXT, and the hypothesis has been put forward by Burbach et al. (1983a, b) that brain OXT may serve as a precursor for behaviorally active neuropeptides which attenuate learning and memory processes more selectively than the nonapeptide or its des-glycinamide-9 fragment. Burbach and De Wied (1981) studied the metabolism of OXT via identification of oligopeptides from OXT as products of proteolytic conversion by brain peptidases. These authors identified the hexapeptide pGlu-Asn-Cys(Cys)-Pro-Leu-GlyNH_2. [pGlu^{4}Cyt^6]-OXT-(4-9) in the incubation media when synaptic membranes were incubated in the presence of OXT. This hexapeptide, which is probably generated by aminopeptidases (Burbach and Lebouille 1983), has been synthesized and investigated for biological activity. This putative cerebral metabolite of OXT was 10^2–10^3 times more effective than the parent nonapeptide in attenuating passive avoidance behavior. Further, it appeared to be more specific toward the central nervous system, since it was free of peripheral agonistic properties in vitro on the isolated uterus of estrogen-pre-

Table 1. Amino acid sequence of oxytocin and fragments

Name	Sequence	Amino acid chain
Oxytocin	Cys-Tyr-Phe-GluNH$_2$-AspNH$_2$-Cys-Pro-Leu-GlyHN$_2$	1–9
Des-GlyNH$_2$-oxytocin	Cys-Tyr-Phe-GluNH$_2$-AspNH$_2$-Cys-Pro-Leu	1–8
Tocinoic acid	Cys-Tyr-Phe-GluNH$_2$-AspNH$_2$-Cys	1–6
pGlu-Oxytocin 4–9	pGluNH$_2$-AspNH$_2$-Cys-Pro-Leu-GlyHN$_2$ Cys	4–9
PLG (MIF-I)	Pro-Leu-GlyHN$_2$	7–9
Pro-Leu	Pro-Leu	7–8
Z-Pro-D-Leu	Z-Pro-D-Leu	7–8
Cyclo-Leu-Gly	Cyclo-Leu-GlyNH$_2$	8–9

treated (sensitized) rats. The uterotonic activity of OXT was not inhibited either by a thousandfold molar excess of the OXT metabolite. This finding suggests that the hexapeptide is also free of peripheral OXT-antagonistic activities. Removal of the C-terminal glycinamide residue from the hexapeptide [pGlu^{4}Cyt^6]-OXT-(4-9) did not destroy the behavioral activity of the metabolite (Burbach et al. 1983 a, b). On the basis of these findings, Burbach et al. (1983 a, b) concluded that proteolytic conversion of OXT by peptidases of brain synaptic membranes may result in the production of metabolites with activity exclusively on the brain, and not in the periphery.

Not only the OXT fragments but also the structural analogs of OXT appeared to be active in various behavioral processes (De Wied and Bohus 1979; Kovács et al. 1982 a). Arginine vasotocin, the oldest neurosecretory principle evolutionarily, was found to attenuate passive avoidance behavior whether injected immediately after the learning trial or shortly before the retention trial. Given before the retention test, arginine vasotocin attenuated passive avoidance as potently as OXT. It is of particular interest that des-glycinamide-9-arginine-vasotocin was only active when it was injected shortly before the retention trial, and not if given after the learning session. While arginine vasotocin contains the N-terminal "ring" structure of OXT together with the C-terminal tripeptide sequence of arginine-vasopressin (AVP), oxypressin contains the "ring" of vasopressin together with PLG as the C-terminal tripeptide. Oxypressin was found to facilitate passive avoidance behavior (De Wied 1976; De Wied and Bohus 1979; Kovács et al. 1982 a), and thus exerted an effect opposite to that induced by OXT on avoidance behavior.

2.5 Local Effects of Oxytocin in Relation to Avoidance Behavior

Only a few data are available in the literature concerning investigations of behavioral effects of OXT following its microinjection locally into brain structures. Such studies are generally performed with the aim of defining the neuroanatomical substrates of the effects of the neuropeptide. It has been reported (Kovács et al. 1979) that the local microinjection of OXT into the dentate gyrus of the dorsal hippocampus and into the mesencephalic dorsal raphe nucleus attenuates passive avoidance behavior. Since the dose of OXT (2.5×10^{-14} mol) was so small as to be insufficient to affect the avoidance response following intracerebroventricular injections, these authors concluded that the hippocampal and midbrain structures might have a role in mediating the effects of OXT on avoidance reactions. Injected into the dorsal septal nucleus, on the other hand, OXT facilitated passive avoidance behavior (Kovács et al. 1979). Interestingly, electrophysiological (Urban 1981; Joëls and Urban 1982) and biochemical (Versteeg 1983; Van Heuven-Nolsen et al. 1984a) evidence also suggests that the dorsal septum might be a brain site responsible for the vasopressin-like effects of OXT (for vasopressin-like and opposite effects, see Table 2).

The physiological significance of these findings is not completely clear at the moment. There are at least three important criteria that have to be fulfilled before it can definitely be concluded that limbic-midbrain structures specifically mediate the influence of OXT on behavioral reactions.

1. The brain structure must be involved in learning and memory processes.
2. The brain structure should contain sensitive receptors that recognize the presence of OXT and mediate its effects.
3. OXT should be transported to these structures by endogenous transport routes.

Although it is premature to draw a final conclusion, the data indicate that the limbic-midbrain structures may fulfil all three criteria, and might thus be regarded as the physiological target sites of OXT in affecting behavior. In accordance with this suggestion, limbic (hippocampus, amygdala, septum) and brainstem structures have been shown to have important roles in various aspects of memory consolidation, storage, and retrieval (Le Piane and Phillips 1978; Destrade and Jafford 1978; McGaugh et al. 1979; etc.). These structures are innervated by extrahypothalamic oxytocinergic fibers, and oxytocinergic synapses terminate there (Sterba 1974; Sterba et al. 1980; Buijs 1978, 1983; Sofroniew 1983; etc.). In addition to the morphological data, there is the fact that a measurable amount of immunoreactive OXT has been shown to be present in the limbic structures of rats (Dogterom et al. 1978; Mens et al. 1982) and mice (Kovács et al. 1985d). Since most of these structures are relatively close to the cerebral ventricles, endogenous OXT normally present in the cereobrospinal fluid (Robinson and Zimmermann 1973; Unger 1977; Mens et al. 1982; Reppert et al. 1984; Robinson and Jones 1982) can likewise reach these putative target sites via diffusion. Since the limbic structures also contain the binding sites for OXT (Ferrier et al. 1983; Brinton et al. 1984; De Kloet et al. 1985), it is likely that local intracerebral microinjection of OXT into limbic-midbrain areas attenuates passive avoidance behavior because the endogenous OXT of these brain areas is also involved in the control of behavioral processes.

Table 2. Effects of oxytocin on behavior: a summary

Behavioral Reaction	Effect of oxytocin	Effect of vasopressin	Species	Route of treatment	Dose of oxytocin (mol/animal)	Reference
Extinction of active avoidance (bench-jumping) reaction	Facilitation	Opposite	Rat	i.p.	2×10^{-11}	SCHULZ et al. (1974, 1976); TELEGDY and KOVÁCS (1979a)
Extinction of active avoidance (pole-jumping) reaction	Facilitation	Opposite	Rat	i.c.v.	10^{-12}	BOHUS et al. (1978b)
Extinction of active avoidance (pole-jumping) behavior	Delay	Similar	Rat	s.c.	2×10^{-10}	WALTER et al. (1978); BOHUS et al. (1978b)
Acquisition of active avoidance (pole-jumping) behavior	Delay	No effect	Rat	i.c.v.	10^{-12}	BOHUS et al. (1978b)
Acquisition of active avoidance (bench-jumping) behavior	No effect	No effect	Rat	i.p.	2×10^{-11}	SCHULZ et al. (1974)
Passive avoidance (step-down) behavior	Attenuation	Opposite	Rat	i.p.	2×10^{-10}	KOVÁCS et al. (1978)
Passive avoidance (step-through) behavior	Attenuation	Opposite	Rat	i.c.v.	10^{-13}	BOHUS et al. (1978a, b); KOVÁCS and DE WIED (1983)
Passive avoidance (step-through) behavior	Attenuation	Opposite	Rat	Hippocampus, dorsal raphe	2.5×10^{-14}	KOVÁCS et al. (1979)
Passive avoidance (step-through) behavior	Facilitation	Similar	Rat	Septum	2.5×10^{-14}	KOVÁCS et al. (1979)
Passive avoidance (choice measure)	No effect	No effect	Rat	?	10^{-8}	SAGHAL and WRIGHT (1984)
Attenuation of puromycin-induced amnesia	No effect	Effective	Mouse	s.c	10^{-7}	WALTER et al. (1975)
Induction of retrograde ammnesia	Effective	No effect	Rat	s.c.	10^{-9}	KOVÁCS and TELEGDY (1985)
Memory recall	Attenuation	Opposite	Human	Intranasal i.v.	$3–5 \times 10^{-8}$	FERRIER et al. (1980) KENNETT et al. (1982); FEHM-WOLSDORF et al. (1984)
Development of acute tolerance to morphine	Attenuation	Not measured	Mouse	s.c	5×10^{-12}	KOVÁCS et al. (1985b)
Development of chronic tolerance to morphine	Attenuation	Not measured	Mouse	s.c. i.c.v. Hippocampus/nucl. accumbens	5×10^{-8} 5×10^{-11} 5×10^{-13}	KOVÁCS et al. (1984a) KOVÁCS et al. (1984a) KOVÁCS et al. 1984a)
Development of dependence on morphine (advanced stage)	Attenuation	Not measured	Mouse	s.c i.c.v. Hippocampus/nucl. accumbens	5×10^{-8} 5×10^{-11} 5×10^{-13}	KOVÁCS et al. (1984a) KOVÁCS et al. (1984a) KOVÁCS et al. (1984a)

on morphine (early stage)						
Acquisition of heroin self-administration in heroin-naive animals	No effect	Attenuation	Rat	s.c.	10^{-9}	VAN REE and De Wied (1977a)
Maintenance of heroin self-administration in heroin-tolerant animals	Attenuation	Not measured	Rat	s.c.	5×10^{-10}	KOVÁCS et al. (1985a); KOVÁCS and TELEGDY (1985)
Development of ethanol tolerance	Attenuation	Not measured	Mouse	s.c.	2.5×10^{-9}	SZABÓ et al. (1983)
Hypothalamic self-stimulation	Increase	Opposite	Rat	i.c.v.	$1.6 \times {}^{-12}$	SCHWARZBERG et al. (1976)
Hippocampal theta activity during paradoxical sleep	Decrease in frequency	Opposite	Rat	i.c.v.	5×10^{-12}	BOHUS et al. (1978b)
Behavioral arousal (scratching, grooming)	Facilitation	Similar	Mouse	i.c.v.	2×10^{-10}	MEISENBERG and SIMMONDS (1982)
Moving, grooming	Facilitation	Similar	Mouse	i.c.v.	10^{-9}	DELANOY et al. (1978)
Scratching	Facilitation	Similar	Rat	i.c.v.	10^{-9}	MEISENBERG (1982)
Induction of barrel rotation	Facilitation	Similar	Rat	i.c.v.	10^{-8}	KRUSE et al. (1977); KORDOWER and BODNAR (1984)
Protection against barrel rotation	Facilitation	Opposite	Rat	i.c.v.	5×10^{-9}	ABOOD et al. (1980)
Induction of full maternal behavior	Facilitation	Similar	Rat	i.c.v.	4×10^{-10}	PEDERSEN et al. (1979), FAHRBACH et al. 1985)
Induction of full maternal behavior	No effect	Not measured	Rat	i.c.v.	4×10^{-10}	RUBIN et al. (1983), SWANSON and BOLWERK (1984)
Apomorphine (low-dose)-induced hypomotility	Attenuation	Not measured	Mouse	s.c.	10^{-7}	KOVÁCS et al. (1985b)
Apomorphine (low-dose)-induced sniffing	Attenuation	Similar	Rat	s.c.	2.5×10^{-8}	XIAO et al. (1984)
Apomorphine (high-dose)-induced hypermotility	Facilitation	Not measured	Mouse	s.c.	5×10^{-13}	KOVÁCS et al. (1985b)
Amphetamine-induced hyper-motility	No effect	Not measured	Mouse	s.c.	5×10^{-8}	KOVÁCS et al. (1985b)
Haloperidol-induced DA receptor supersensitivity	Attenuation	Not measured	Mouse	s.c.	5×10^{-10}	KOVÁCS et al. (1985b)
Rotational (turning) behavior after unilateral lesion of the substantia nigra	Ipsilateral turning	Similar	Rat	i.c.v.	5×10^{-11}	SCHULZ et al. (1979)

In some publications the dose of oxytocin can only be found in international units. The assumption has been made in these cases that the activity was 300 IU/mg and this value was used to calculate the dose in moles per animal

2.6 Endogenous Oxytocin and Avoidance Behavior: Passive Immunization with Antioxytocin Serum

The principle of passive immunization has been successfully used in behavioral experiments to study the physiological importance of neurohypophyseal neuropeptides in brain functions (Van Wimersma Greidanus et al. 1975; Bohus et al. 1978b; Kovács et al. 1980, 1982b; Kovács and Telegdy 1983). The basic idea is that the intracerebroventricular injection of specific anti-OXT serum neutralizes endogenous OXT in the cerebral ventricles and also penetrates the central nervous system, suggesting that brain OXT is also being neutralized. Early data relating to anti-OXT serum were negative. Van Wimersma Greidanus et al. (1975) found that the intracerebroventricular injection of anti-OXT serum did not affect the passive avoidance response in rats. In this experiment, however, anti-OXT serum was used as a control to prove the specificity of other antisera, and thus the control animals (treated with normal rabbit serum) were trained to display maximum passive avoidance. A further improvement due to the central injection of anti-OXT serum was not possible. Realizing this, the same group of investigators (Bohus et al. 1978b) injected 1 μl undiluted anti-OXT serum into the lateral ventricle and investigated the avoidance in a control group, which exhibited moderate avoidance only. Under these circumstances, anti-OXT serum resulted in a significant delay of the extinction of a pole-jumping avoidance reaction. This effect of the antiserum was exactly the opposite of that induced by peripheral (Schulz et al. 1974, 1976; Telegdy and Kovács 1979a, b; Kovács and Telegdy 1982) or central (Bohus et al. 1978a, b; Kovács et al. 1982a) OXT treatment. Injected immediately after the learning trial of a one-trial-learning passive avoidance reaction experiment, the antiserum improved passive avoidance behavior in 24- or 48-h retention sessions (Bohus et al. 1978b), whereas OXT under identical experimental conditions attenuated passive avoidance (Bohus et al. 1978a, b, Kovács et al. 1982a; Kovács and De Wied 1983). Probably in connection with changes in avoidance behavior, the administration of anti-OXT serum and OXT caused opposite changes in the power spectrum of hippocampal theta activity during paradoxical sleep (Bohus et al. 1978b; Urban 1981). Although anti-OXT serum did not affect the total number of paradoxical sleep episodes, the antiserum resulted in an increased proportion of the higher theta frequencies and a decreased number of the low-frequency components. Besides behavioral and bioelectric changes, central treatment with anti-OXT serum also affected dopaminergic neurotransmission in a direction opposite to that caused by OXT itself (Kovács and Telegdy 1983). Since the neutralization of OXT by the intracerebroventricular injection of anti-OXT serum resulted in a facilitation of passive avoidance behavior, resistance to extinction in the active avoidance paradigm, and an increase in the activity of the limbic theta-generating and brainstem dopaminergic cells (Bohus et al. 1978b; Kovács and Telegdy 1983), it is conceivable that the neuronal substrate of these processes is under the tonic control of endogenous OXT.

3 Oxytocin and Narcotic Addiction

3.1 Oxytocin and the Development of Morphine Tolerance/Dependence

In addition to the proposed effects of OXT on fear-motivated avoidance behavior, it has become increasingly evident that the neuropeptide and its fragments are implicated in the development of tolerance to and dependence on narcotic analgesics, such as morphine or heroin. In general, the data suggest that OXT does not interfere with the acute effects of narcotic analgesics, but modifies the adaptive response of the organism that normally develops with repeated drug exposure (Kovács et al. 1984a, 1985b; Kovács and Telegdy 1985). This phenomenon, called "tolerance", is manifested by a diminished response to a drug after prior exposure. Their antinociceptive effect is the most frequently used test model in the case of narcotic analgesics, and tolerance appears here as a decrease in the antinociceptive response. Since adaptation and learning are likely to be involved in the cellular events leading to tolerance (Himmelsbach 1943; Siegel 1975, 1976, 1978; Wikler 1980), and since OXT attenuates learning and memory processes, the question logically arose whether OXT is able to influence the development of acute and chronic tolerance. Van Ree and De Wied (1976) reported that the subcutaneous injection of OXT facilitated the development of tolerance to morphine in male rats. OXT was given in this experiment 1 h prior to repeated morphine injections and tolerance to the antinociceptive effect was tested on a hot plate (54 °C). The interpretation of these data is somewhat complicated by the fact that morphine and naloxone were coadministered on the test day when analgesia was measured, and also by the high dose of morphine (80 mg/kg) that induced maximum analgesia in the controls. Schmidt et al. (1978), on the other hand, failed to observe any significant effect of OXT on the tolerance to and dependence on morphine in mice. It is of interest to note that vasopressin and its behaviorally active fragments facilitated learning/memory processes (De Wied et al. 1976; Bohus et al. 1978a; etc.) and, in complete agreement with these findings, the peptide also facilitated the development of morphine tolerance in mice (Krivoy et al. 1974), rats (Van Ree and De Wied 1976, 1977a), and cats (Cools et al. 1977) and restored the development of diminished tolerance in rats with diabetes insipidus (De Wied and Gispen 1976) or that induced by central treatment with antivasopressin serum (Van Wimersma Greidanus et al. 1980). Besides the differences in experimental protocol, the dose of OXT might also be an essential factor in explaining the contradictory findings from the various laboratories. Kovács et al. (1985b) therefore investigated the effects of wide dose ranges of OXT on the development of tolerance to the antinociceptive effect of morphine in pellet-implanted mice. OXT was injected subcutaneously 2 h prior to the implantation of the morphine pellet. OXT dose-dependently inhibited the development of tolerance to morphine. However, high doses (5×10^{-8}–10^{-7} mol/mouse) were required to attenuate this type of chronic tolerance. Much lower doses (10^4 times lower) of OXT inhibited the development of acute morphine tolerance induced by a single injection of a high dose of morphine and tested via the antinociceptive effect of a second injection of a small dose of morphine (Kovács et al. 1985b). The reason why OXT appears to have considerably more effect on the develop-

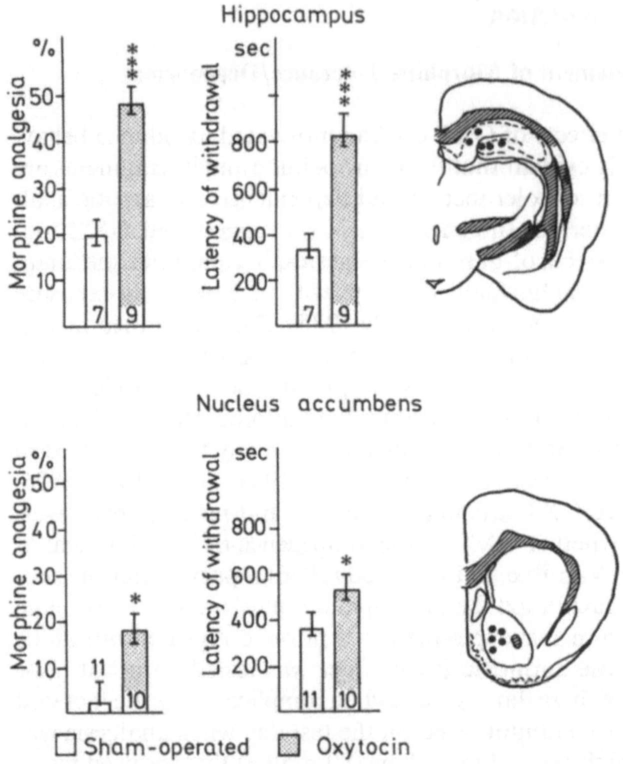

Fig. 1. Limbic and mesolimbic brain structures and the effect of oxytocin on morphine tolerance and withdrawal. Morphine analgesia: Mice were rendered tolerant/dependent on morphine by subcutaneous implantation of a morphine pellet (for details, see Kovács et al. 1984 a). Tolerance was measured 3 days after pellet implantation: the tail flick latencies were measured on the day before and 30 min after the injection of a test dose of 5 mg/kg morphine and the analgesia was expressed as a percentage of the pretreatment value. Tolerance in the sham-operated animals appears as a low percentage value of morphine analgesia. The increase in morphine analgesia in mice with microinjection of OXT into the hippocampus or the nucleus accumbens indicates that OXT interferes with the development of morphine tolerance. Latency of withdrawal: Naloxone (1 mg/kg) was injected intracerebroventricular into the tolerant/dependent mice and the latency of stereotyped withdrawal jumpings was measured. Following injection, 5×10^{-11} mol OXT was required to attenuate morphine tolerance/dependence and a dose of 5×10^{-12} mol was insufficient. The *bars* indicate the local effect of 5×10^{-13} mol OXT microinjected into limbic-mesolimbic brain structures. These findings suggest a preferential sensitivity of the hippocampus and the nucleus accumbens toward exogenously administered OXT. *Full circles* indicate the typical localization of the tip of the cannula used for microinjection. ***$p<0.001$, *$p<0.05$ vs sham-operated group. The *numbers in the bars* indicate the number of mice

ment of acute than of chronic tolerance is poorly understood. Recently, Ritzmann et al. (1984) concluded that the pellet implantation technique is an experimental paradigm that can be expected to produce circumstances during which environment-independent components of functional morphine tolerance would contribute largely to the diminished response to analgesics. In contrast, environmental cues probably play a larger role in acute tolerance, in which repeated drug injections and testings were performed in the same environmental situation. Thus acute tolerance is more likely to be related to environment-dependent functional tolerance. If this hypothesis, which needs further experimental support, proves to be true, one could conclude that OXT more effectively attenuates environment-dependent components of tolerance than environment-independent ones. Another equally likely explanation, however, is that acute and chronic tolerance represent sequential stages of a common process in tolerance development (Way and Rezvani 1984), and that the early stages (i.e., acute tolerance) are more susceptible to OXT.

Repeated exposure to narcotic analgesics frequently results in the development of physical dependence on the drug. Classical theories of drug addiction have therefore considered tolerance and physical dependence to be two phenomena of one and the same mechanism. One of the main arguments in favor of this view is that there is no known instance of physical dependence occurring in the absence of tolerance. Recent observations, however, suggest that the development of tolerance can be manipulated without affecting the degree of physical dependence (Tabakoff and Ritzmann 1977; Ritzmann et al. 1984). Accordingly, it was of interest to test whether OXT also modifies the development of physical dependence on morphine.

Van Ree and De Wied (1976) investigated the influence of OXT on the development of dependence in female rats, using multiple morphine injections for 3 days. Naloxone resulted in a loss of body weight and a decrease in body temperature in rats that received OXT before morphine injections. These authors interpreted their findings as indicating a facilitation of morphine dependence by OXT. This effect, however, probably reflects the influence of OXT on a marginal degree of physical dependence, since the control female rats did not lose body weight and did not exhibit hypothermia following naloxone. In an advanced stage of physical dependence, the situation might be different, in mice in particular. A single subcutaneous injection of OXT given shortly before implantation of the morphine pellet resulted in a delayed appearance of naloxone-precipitated withdrawal jumpings, a withdrawal symptom likely to be related to the functional imbalance of central dopaminergic neurons. In addition to the withdrawal jumpings, OXT also diminished other symptoms of the precipitated withdrawal syndrome (e.g., piloerection, salivation, tremor) (Kovács et al. 1984b, c).

It appears probable that the effects of OXT on the development of tolerance to and dependence on morphine are mediated by central nervous structures (Fig. 1). In support of this idea, OXT was 10^2–10^3 times more potent following intracerebroventricular injection than after subcutaneous administration (Kovács et al. 1985b). Since the dispositional tolerance (tolerance due to altered tissue distribution, metabolism, or elimination) largely depends on peripheral variables and OXT does not affect the brain levels of morphine (Schmidt et al. 1978), a central site of action strongly suggests that OXT attenuates adaptive mechanisms normally involved in the development of functional tolerance.

3.2 Oxytocin and Reinforcement

Oxytocin is effective in modifying central nervous system mechanisms underlying reinforcement, and mainly those related to drug reinforcement and electrical self-stimulation. Drug reinforcement is one of a large class of reinforcing events (e.g., food and water presentation to deprived individuals, the opportunity to engage in sexual behavior, or – under experimental circumstances – electrical stimulation of the brain). Heroin was used as reinforcer in most of the studies relating to the effect of OXT and its fragments on drug addiction (Van Ree and De Wied 1977a; Kovács et al. 1985a; Kovács and Telegdy 1984, 1985; Van Ree 1982). Depending on the stage of heroin addiction, the effects of OXT proved to be different. The acquisition of heroin self-administration was not influenced by OXT in rats (Van Ree and De Wied 1977a). Thus the subcutaneous injection of 10^{-9} mol OXT did not interfere with the number of heroin-reinforced bar pressings. However, there was a tendency to increase heroin self-administration following OXT (Van Ree and De Wied 1977a), though this did not reach a level of statistical significance when compared to vehicle-treated controls. Neutralization of OXT in the central nervous system and in the cerebrospinal genid by the intracerebroventricular injection of anti-OXT serum likewise proved to be ineffective in modulating heroin self-administration (Van Ree and De Wied 1977b). The situation is completely different when the action of OXT on the maintenance of heroin self-administration is studied in rats previously rendered tolerant to and partly dependent on heroin (Kovács and Telegdy 1984, 1985; Kovács et al. 1985a). Once a stable level of heroin self-administration was achieved in these addicted rats, the subcutaneous injection of OXT (5×10^{-10}–5×10^{-9} mol/animal) dose-dependently decreased the amount of heroin administered. The effect was maximal 1 h after peptide treatment. Self-administration procedures have been used as a model for measurement of the reinforcing efficacy of drugs (Thompson and Pickens 1975), and it appears likely that OXT decreases the reinforcing efficacy of heroin in the heroin-tolerant organism by inducing a shift in the degree of tolerance/dependence (Kovács and Van Ree, in preparation).

In recent years, some results have led to the conclusion that neural processes mediating the positively reinforcing characteristics (e.g., euphoria) of narcotic drugs might be related to mechanisms through which narcotics affect electrical self-stimulation. This hypothesis (Maroli et al. 1978; Esposito and Kornetsky 1977) implies that drug-induced reinforcement and electrically induced reinforcement (self-stimulation) are related to a final common reinforcing pathway in the brain. As far as the effects of OXT are concerned, however, the situation seems to be more complex. Whereas OXT decreases the self-administration of heroin in the tolerant organism (Kovács et al. 1985a), electrical self-stimulation behavior, elicited by suprathreshold high-frequency (90–130 cps) stimulation of the medial forebrain bundle, was significantly increased by intracerebroventricular OXT treatment (Schwarzberg et al. 1976).

The data reviewed in this section suggest that the reinforcement (reward) mechanism triggered by heroin and involved in the maintenance of drug-seeking behavior in the heroin-addicted organism is under the control of the pituitary neuropeptide OXT. Under physiological circumstances, OXT probably does not affect drug-seeking behavior alone, but in concert with vasopressin (Van Ree and De Wied 1977a, b; Van Ree 1982) and prolactin (Drago et al. 1984, 1985), and a disturbance in one or more of these systems may lead to a state in which drug-seeking behavior can be expected to develop.

3.3 Oxytocin Fragments and Narcotic Addiction

Various fragments of OXT have been investigated on narcotic addiction and there is no consensus concerning the effects, potencies, or modes of action of these fragments (for recent reviews, see Van Ree 1982, 1983; Ritzmann et al. 1984). Des-glycinamide-9-OXT, like the hormonally active nonapeptide, decreases heroin self-administration in heroin-addicted rats (Kovács and Van Ree, in preparation), suggesting that the C-terminal glycinamide and the hormonal activity are not essential for this action. The most contradictory published data concern the effects of the C-terminal tripeptide of OXT. It has been reported that PLG facilitates the development of tolerance to morphine in male rats and the dependence on morphine in female rats (Van Ree and De Wied 1976). Others (Ritzmann et al. 1980) found that this tripeptide inhibited rather than facilitated the development of morphine tolerance. These latter studies mostly used higher peptide doses and mice as test subjects. The discrepancy may result in part from the experimental protocol, as indicated by the findings of Székely et al. (1979), who found that PLG facilitated or did not affect morphine tolerance in rats, depending on the interval between peptide treatment and morphine challenge. The picture is even more complex with withdrawal, since the tripeptide affects some but not all symptoms of morphine withdrawal (Székely et al. 1979). According to Kastin et al. (1980), the tripeptide also interacts with endogenous opiate binding sites, which effect would easily explain its interactions with morphine tolerance/dependence. Others (Dickinson and Slater 1980; Conteras and Takemori 1981; Bhargava 1983) have concluded, however, that there is no evidence that PLG affects morphine-induced analgesia or the binding sites labeled by various ligands for opiate receptors.

Besides PLG, the cyclic C-terminal dipeptide (cyclo-Leu-Gly) and the N-terminal dipeptide of PLG (Pro-Leu, and mainly its protected analog, Z-Pro-D-Leu) have been reported to affect narcotic tolerance and dependence (Ritzmann et al. 1980, 1984; Bhargava et al. 1983; Kovács et al. 1981, 1984a, c, 1985b). Z-Pro-D-Leu dose-dependently inhibits the development of tolerance to morphine following subctuaneous, intracerebroventricular, or intracerebral injections (Kovács et al. 1984a, 1985b), and in some respects the effects of the dipeptide resemble those of central OXT treatment. In addition to tolerance and dependence, Z-Pro-D-Leu also reduces the amount of self-injected heroin administered by heroin-addicted rats (Kovács and Telegdy 1984, 1985; Kovács et al. 1985a). These dipeptides derived from PLG do not affect morphine-induced analgesia (Kovács et al. 1981; Bhargava et al. 1983) or the binding of labeled naloxone to brain opiate receptors (Bhargava et al. 1983). As to a possible mode of action, these dipeptides profoundly affect the dopaminergic neurotransmission (Kovács et al. 1981, 1983b, 1984a; Ritzmann et al. 1984; Bhargava 1983), and dopamine receptor blockers interfere with their effects on narcotic addiction (Kovács et al. 1984c and f).

3.4 Endogenous Oxytocin and Morphine Tolerance/Dependence

The idea has recently been put forward (Kovács et al. 1985d) that endogenous OXT in the limbic areas might be involved in the development of narcotic tolerance/dependence. In agreement with this hypothesis, it has been found that the immunoreactive OXT content of the mouse hippocampal tissue is increased by the acute injection of an analgesic dose of morphine (Kovács et al. 1985d). This effect of the narcotic analgesic was reversed by naloxone pretreatment, indicating the specific involvement of endogenous opiate receptors. It was assumed (Kovács et al. 1985d) that the increased OXT level in the hippocampal formation following morphine treatment reflects an increased accumulation of OXT in the OXT-containing nerve fibers and endings in the hippocampus (Buijs 1978, 1983; etc.), and hence the release of this neuropeptide might be decreased by acute morphine challenge. Consequently, the availability of OXT at the functionally distal OXT-sensitive cells would be decreased. Other possibilities, however, such as an altered biotransformation and synthesis or changes in the binding of endogenous OXT to the limbic receptors (Ferrier et al. 1983; De Kloet et al. 1984), should also be taken into account. In the light of the studies employing exogenous OXT injections in the hippocampus, which resulted in an attenuation of morphine tolerance/dependence, a decreased availability of endogenous OXT would be consistent with and in turn could contribute to the development of tolerance to morphine.

In contrast to the situation following acute morphine injection, the hippocampal OXT level appeared to be low in tolerant/dependent animals. This finding could be related to the chronic presence of morphine and might be associated with a decreased activity of limbic oxytocinergic fibers. During the precipitated withdrawal syndrome, i.e., when injection of the opiate receptor blocker resulted in a rapid onset of stereotyped jumpings (and of salivation, tremor, ptosis, diarrhea, piloerection, etc.), the OXT level in the hippocampus was abruptly elevated. Whether this sudden increase was related to an increased synthesis and availability of OXT or, conversely, to an even more pronounced intraneural accumulation associated with a decreased availability, is a matter of speculation and cannot be decided before measurement of the OXT turnover.

Taken together, the data raise the possibility that acute morphine injection and the development of morphine tolerance/dependence are closely associated with changes in the level of endogenous OXT in the mouse hippocampal formation.

3.5 Oxytocin and Ethanol Tolerance

Considerably fewer data are available on the effects of OXT on the chronic effects of alcohol (for recent reviews, see Ritzmann et al. 1984; Rigter et al. 1980). Treatment with OXT during induction of tolerance did not affect the tolerance to the hypothermic effect of ethanol in ethanol-fed C57BL/6 mice (Hoffmann et al. 1978). In CFLP mice, on the other hand, OXT inhibited the development of tolerance to the hypothermic action of force-injected ethanol (Szabó et al. 1983,

1984a). Not only the nonapeptide but also the C-terminal fragments (PLG, Z-Pro-D-Leu) inhibited ethanol tolerance in CFLP mice (Szabó et al. 1985). Since blood alcohol levels were not affected by OXT (Hoffmann et al. 1978), one might assume that OXT interferes with the development of functional tolerance to ethanol (Szabó et al. 1983, 1984).

4 Oxytocin in Maternal and Sexual Behavior

Pedersen and Prange (1979) conducted the first studies to determine whether OXT, a peptide that produces uterine contractions and milk ejection and, additionally, is native to the brain and the posterior pituitary gland, plays a role in the induction of maternal behavior. These authors observed that, following the intracerebroventricular injection of a relatively high dose of OXT (4×10^{-4} mol), almost 50% of intact virgin female rats displayed full maternal behavior toward foster pups. To be rated as fully maternal, the animals had to display all the behavioral characteristics of the nursing rat (grouping, licking, crouching, retrieving the pups, and nest building). Interestingly, a partial maternal response was seldom produced by OXT, and thus the effect was of an all-or-none nature. Data pertaining to estrus cycle strongly suggested that the maternal response to OXT was partly estrogen-dependent (Pedersen and Prange 1979). Estrogen dependency was demonstrated by a high rate of full maternal response after OXT in ovariectomized estrogen-primed females, as compared to the virtual absence of response in ovariectomized animals not primed with estrogens. In a follow-up study, Pederson et al. (1982) reported that the effect of OXT on maternal behavior was dose-dependent and reasonably specific. Of the 20 hormones and hormonal fragments tested, only tocinoic acid, the ring structure of OXT, exhibited a maternal effect comparable to that of OXT. The C-terminal tripeptide of OXT appeared to be ineffective. Although the first data indicated that vasopressin did not induce a maternal response (Pedersen and Prange 1979), the difference between OXT and vasopressin turned out to be quantitative and not qualitative (Pedersen et al. 1982). Recently, Pedersen et al. (1984) investigated the physiological significance of endogenous OXT in maternal behavior. Nulliparous ovariectomized female rats were treated with a combination of estrogen and progesterone, which induced a high incidence of maternal behavior. Subsequently, anti-OXT serum was injected (intracerebroventricularly) and a lower cumulative incidence of full maternal behavior was observed. Antivasopressin and antiprolactin sera were ineffective in the same experimental paradigm.

The different approach used by Fahrbach et al. (1985) resulted in similar conclusions as regards the role of OXT in maternal behavior. These investigators used pregnancy-terminated rats (pregnant rats hysterectomized, ovariectomized, and injected with estradiol), in which there is a high incidence of maternal behavior in response to foster pups. Fahrbach et al. (1985) reported that the central administration of $d(CH_2)_5$-ornithine-8-vasotocin, an analog antagonist of OXT, completely disrupted maternal behavior in these animals. The OXT antagonist also delayed the onset of maternal behavior in lactating rats (Fahrbach et al. 1985).

While the clearest effects of OXT on full maternal behavior were described in Sprague-Dawley rats of the Zivic Miller Laboratories (Pedersen and Prange 1979; Pedersen et al. 1982; Fahrbach et al. 1985), others were unable to reproduce these findings in estrogen-primed ovariectomized virgin female Wistar or Sprague-Dawley rats (Rubin et al. 1983; Swanson and Bolwerk 1984). In addition to the possible involvement of strain differences, Fahrbach et al. (1985) offer an alternative explanation for the contradictory results. They found that when the effect of central OXT treatment was studied in animals habituated to their home cages, the peptide had no effect on maternal behavior. In a novel environment, on the other hand, OXT potently facilitated maternal behavior, suggesting that the novelty of or the familiarity with the environment can influence the display of maternal behavior and modify the action of OXT.

Although further research is necessary for an understanding of the physiological role of OXT in the onset of maternal behavior, the initial data are encouraging, especially as these are the first reports in which peptide manipulations have been shown to turn a virgin rat into a maternal one. Since OXT is released during parturition (Chard et al. 1971) and nursing (Folley and Knaggs 1966), it seems possible that OXT normally triggers the onset of a maternal response toward the offspring.

Much less is known about the putative role of OXT in sexual behavior. Vaginal distension is well known to release OXT from pituitary stores (Ferguson's reflex) and to induce sexual receptivity in females (Tindal 1974). The only report available so far (Caldwell et al. 1984) indicates that intracerebroventricular OXT treatment facilitates female lordosis behavior, a characteristic component of female sexual behavior, in estrogen-primed ovariectomized rats.

5 Oxytocin and Behavioral Arousal

Central (intracerebroventricular) treatment of rats and mice with high doses of peptides induces behavioral arousal characterized by ataxia, opisthotonus, prostration, and barrel rotation (unusual rotational behavior along the sagital body axis). It has been shown that, like other peptides (vasopressin, somatostatin), OXT too can promote these behavioral symptoms (Kruse et al. 1977; Kordower and Bodnar 1984). The ED_{50} of central OXT in producing these supposedly toxic effects is estimated to be 7×10^{-9} mol (Kruse et al. 1977) and, as far as doses are concerned, OXT was 30–40 times less effective than argininevasopressin or vasotocin. Abood et al. (1980) confirmed that vasopressin (and somatostatin) induced barrel rotation, but did not observe these symptoms following intracerebroventricular treatment with very high doses (2×10^{-8} mol) of OXT. In contrast, 1–5×10^{-9} mol OXT, injected intracerebroventricularly shortly prior to the administration of AVP, protected against AVP-induced barrel rotation. This protective effect of OXT was rather selective, since other peptides (somatostatin, luteinizing-hormone-releasing hormone, and Gly-Leu-Tyr) were unable to prevent AVP-induced barrel rotation (Abood et al. 1980). Pointing to the most likely brain sites of the reported protective effect of OXT, Abood et al. (1980) established that lesioning of the dorsal hippocampus prevented barrel rotation elicited by AVP. Thus hippocampal lesions and OXT treatment exert similar effects.

In mice, Delanoy et al. (1978) found that intracerebroventricular injection of a high dose (10^{-9} mol) of OXT caused a behavioral syndrome characterized by grooming, moving, foraging, and squeaking. These behavioral symptoms, however, were not specific to OXT, since LVP, AVP, and vasotocin appeared to be active as well. The jerky behavioral changes were on no occasions associated with electrical seizures in the hippocampus, amygdala, or cerebral cortex. According to Meisenberg (1981), scratching rather than grooming is the predominant symptom of behavioral arousal induced by OXT in mice. Other symptoms more frequently observed by this author were squeaking, foraging through wood shavings, occasional jumpings, and head-to-tail circling. The dose of OXT required to precipitate these effects was 10^{-9}–3×10^{-8} mol intracerebroventricularly, with the exception of scratching, which frequently occurred following 5×10^{-12}–2×10^{-11} mol OXT. Although these behavioral reactions were regarded as toxic side effects (Kruse et al. 1977; Cohn and Cohn 1975) of peptide treatment, or at best the pharmacological consequences of an overdose of peptides (Delanoy et al. 1978), it is of interest that environmental cues profoundly modify these symptoms. Accordingly, when mice are pushed into a Plexiglas tube ("tube test" according to Meisenberg 1981), OXT does not induce grooming and scratching, but dose-dependently stimulates the escape-directed activity of the animals.

In a follow-up experiment designed to investigate the structural requirements of peptides to induce behavioral arousal, Meisenberg and Simmonds (1982) found that des-glycinamide-9-OXT was considerably less active than the parent nonapeptide. Thus the structure-activity requirements were completely different from those observed in fear-motivated avoidance behavior (De Wied 1976; Kovács et al. 1982a) or narcotic addiction (Van Ree 1982; Kovács and Van Ree, in preparation). In general, the activity of neurohypophyseal peptides and analogs in producing behavioral arousal was parallel to the vasoconstrictor effects of these peptides (Meisenberg and Simmonds 1982). Despite the parallelism, Meisenberg (1982) concluded that vasoconstriction was not essential for these effects.

Meisenberg (1981) also found that intracerebroventricular OXT treatment reduced the period of tonic immobility in a mouse swim test (behavioral despair test), in which antidepressants and psychostimulants are known to reduce immobility (Porsolt et al. 1977). It is not clear whether this effect of OXT is specific, or a part of behavioral arousal.

In contrast to the effects of OXT on fear-motivated behavior, where a single injection of the peptide can affect subsequent behavior for hours or days, behavioral arousal is a snydrome with a short-term nature (Meisenberg 1981, 1982; Kruse et al. 1977). The short duration probably arises because the symptoms depend on the presence of a high concentration of OXT in the brain or brain ventricles. In support of this notion, Meisenberg and Simmonds (1984a) reported that the effect of OXT on behavioral arousal was potentiated by amastatin, a drug known to inhibit several aminopeptidases in the brain. However, using structural analogs of OXT, Meisenberg and Simmonds (1984b) found no clear correlation between the chemical structure and the duration of action.

In conclusion, the biological significance of behavioral arousal following central treatment with high doses of OXT is not clear. Some authors (Meisenberg and Simmonds 1984 a, b) view this behavioral reaction as a model for understanding of the physiological functions of neuropeptides in the brain. The major difference between OXT-induced behavioral arousal and the putative physiological effects of OXT (e.g., on fear-motivated avoidance behavior and narcotic addiction) is that behavioral arousal can be observed following central treatment exclusively and is precipitated only by very high doses of OXT. This effect is of short duration and probably lacks specificity.

6 Oxytocin as Neuromodulator: A Way to Influence Behavior

The existence of an oxytocinergic neuronal network in the brain (Sterba 1974; Buijs 1983; Kozlowski et al. 1983; Sofroniew 1983), together with putative OXT-sensitive receptors in the brain (Ferrier et al. 1983; Brinton et al. 1984; De Kloet et al. 1984), has led to the hypothesis that OXT may act as either a neurotransmitter or a neuromodulator in the central nervous system. That OXT is released in the brain by depolarizing stimuli (Buijs and Van Heerikhuize 1982) argues for a transmitter function, as reviewed recently by Buijs (1983). As a neuromodulator, OXT would be expected to modify ongoing neuronal processes. In fact, it has been shown that OXT modulates (amplifies or dampens) the neuronal activity of various transmitter pathways in the brain (Telegdy and Kovács 1979 a, b; Kovács and Telegdy 1982, 1984 a; Schwarzberg et al. 1981, 1984; Van Heuven-Nolsen et al. 1984 a). The high amount of immunoreactive OXT in brain regions which contain cell bodies of various dopaminergic, noradrenergic, and serotoninergic pathways provides an anatomical background for the modulatory function (Hawthorn et al. 1984). However, a rigid distinction between modulatory and transmitter functions may not be necessary in the light of the most recent functional and morphological data.

6.1 Oxytocin and Brain Dopamine

6.1.1 Effects of Oxytocin on Brain Dopamine Metabolism

Telegdy and Kovács (1979 a, b) studied the effects of systemically administered OXT on the steady state level and α-methyl-p-tyrosine-methylester-HCL (α-MPT)-induced disappearance (utilization) of dopamine (DA) in the rat brain. These authors found that DA utilization was decreased by OXT in the striatum, whereas hypothalamic DA utilization was increased after intraperitoneal OXT treatment. Schwarzberg et al. (1981) investigated the effects of central OXT treatment on brain DA, and found that intracerebroventricular OXT decreased the endogenous levels of DA in the mesencephalon and the hypothalamus. They concluded that OXT may induce changes in dopaminergic neurotransmission which outlast the actual presence of the injected peptide in the ventricular system. Kovács and Telegdy (1983) correlated OXT-induced changes in dopaminergic neurotransmission with those induced by this neuropeptide in avoidance behavior. The working hypothesis was that the biochemical changes most likely to be causally involved in the effects on avoidance behavior are those which follow the same pattern of changes as avoidance behavior itself. Since both OXT and the behaviorally active fragment des-glycinamide-9-OXT attenuate avoidance behavior following peripheral or intracerebroventricular administration (see above), and since intracerebroventricular injection of anti-OXT serum exerts opposite effects on behavior, Kovács and Telegdy (1983) used these criteria to correlate the effects of OXT on DA and behavior. It was found that subcutaneous and intracerebroventricular injection of OXT decreased the utilization of DA in the mesencephalon. Des-glycinamide-9-OXT had the same effect, while anti-OXT serum resulted in a facilitation of DA utilization (Kovács and Telegdy 1983). Thus it was concluded that the DA turnover in the mesencephalon, which contains the dopaminergic cell bodies of both major ascending dopaminergic pathways (nigrostriatal in the substantia nigra and mesolimbic in the ventral tegmental area: A_9 and A_{10} regions respectively), is closely associated with the effects of OXT and its behaviorally active derivatives on behavioral reactions. In the light of the recent morphological findings on the descending OXT pathways (Swanson and Sawchenko 1983) and the relatively high OXT content of DA-containing mesencephalic nuclei (Hawthorn et al. 1984), the modulatory role of endogenous OXT on brainstem DA utilization (Kovács and Telegdy 1983) might be of functional significance for the organization of behavior.

The detailed mechanism of interaction of OXT with brain DA is not known. In a recent study, Van Heuven-Nolsen et al. (1984) investigated the effects of graded doses of OXT on DA utilization in microdissected brain nuclei, using a sensitive radiochemical assay. In general, the effects of OXT on DA utilization were modest in terms of brain regions where OXT was active as compared to the large number of nonresponsive regions. Probably due to the effort of the authors to dissect as small as possible brain nuclei in order to stay within specific dopaminergic projections, the α-MPT-induced DA decrease resulted in DA levels close to the detection limit of the assay in some of the nuclei. Hence changes of the magnitude of 30%–50% did not appear to be significant in these nuclei (e.g., in the substantia nigra and in the basal amygdaloid nucleus). For example, in the mesencephalic ventral tegmental area (A_{10} region: Björklund et al. 1975), which contains the cell bodies of the ascending mesolimbic DA fibers innervating the nucleus accumbens, forebrain, and amygdala, OXT caused a decrease of 38% in DA utilization (Van Heuven-Nolsen et al. 1984a). This effect, which was not significant due to the relatively large spreading, is indicative of changes of DA turnover in the same direction and of the same magnitude as the significant reduction in DA utilization reported for the mesencephalon as a whole (Kovács and Telegdy 1983). Further, OXT also decreased the utilization of DA in the medial septal nucleus, another brain nucleus innervated by the mesolimbic dopaminergic projection (Björklund et al. 1975).

In the most characteristic terminals of the mesolimbic DA pathway (nucleus accumbens, amygdaloid nuclei), Van Heuven-Nolsen et al. (1984a) found no changes in DA utilization following a wide range of doses of intracerebroventricularly administered OXT. In the mouse brain, however, the subcutaneous injection of OXT decreased the α-MPT-induced utilization of DA in the nucleus accumbens-septal complex (Kovács et al. 1984c). In the same region, chronic OXT treatment significantly decreased the number of apparent binding sites of DA, as measured by the binding of ^{3}H-spiroperidol. These findings indicate that OXT modulates the activity of mesolimbic dopaminergic neurons. The functional significance of this modulatory action is discussed in Sect. 6.1.2.

In addition to the mesolimbic dopaminergic projection, OXT also appeared to affect DA utilization in the basal ganglia, innervated mainly by the nigrostriatal DA pathway. In fact, the early findings (Telegdy et al. 1979a, b) suggested that the opposite effects of OXT and vasopressin on avoidance behavior could be related to an opposite influence of these peptides on DA utilization in the striatum. This hypothesis, which has recently been extended by others (Van Heuven-Nolsen and Versteeg 1984), was mainly based on the observation that OXT decreased, while vasopressin increased, the disappearance of striatal DA following synthesis inhibition with α-MPT (Kovács et al. 1977; Tanaka et al. 1977; Telegdy and Kovács 1979a, b; Van Heuven-Nolsen and Versteeg 1984). However, the decrease in striatal DA utilization was only observed when the neuropeptide was coadministered with the synthesis inhibitor (Telegdy and Kovács 1979a, b; Kovács and Telegdy 1983). If the neuropeptide was administered after complete inhibition of tyrosine hydroxylase had been achieved with α-MPT, the neuropeptide also facilitated DA utilization. These findings suggest that, besides an action of OXT on the striatal turnover of DA, OXT can also modify the synthesis of DA (Kovács and Telegdy 1983) in the striatum. In the globus pallidus OXT decreased the utilization of DA (Van Heuven-Nolsen et al. 1984a).

6.1.2 Oxytocin and Dopamine-Mediated Drug Responses

Certain drugs affect dopaminergic neurotransmission, which in turn has the consequence of altering the animal's activity, producing stereotyped reactions (sniffing, grooming) or inducing circling (turning) behavior. Such drugs include amphetamine, which is known to release DA from dopaminergic terminals, and apomorphine, which can stimulate either pre- or postsynaptic receptors, depending on the dose of the receptor stimulant. DA receptor blockers (e.g., haloperidol) are also frequently used to induce DA receptor supersensitivity following prolonged treatment. The changes induced by these drugs in the locomotor activity of the animals are most probably mediated by the mesolimbic dopaminergic projection (ventral tegmental area to the nucleus accumbens and to other limbic structures), whereas stereotyped behavior or turning following unilateral destruction of the substantia nigra are the results of an altered DA transmission in the nigrostriatal projection (substantia nigra to the nucleus caudatus) (for surveys, see Costall et al. 1984; Anden et al. 1984). Neuropeptides can inhibit or potentiate these drug effects, and their interactions can be used to characterize the mode of action and site of action. Peripheral injection of graded doses of OXT did not affect the motor hyperactivity induced by a low dose of amphetamine in mice (Kovács et al. 1985b), suggesting that OXT did not influence the release of DA in the mesolimbic DA pathway. Biochemical data pointing to the lack of effect of OXT on the in vitro release of ^{3}H-DA support this view (Kovács et al. 1984c). Apomorphine was used to characterize possible interactions of OXT with pre- and postsynaptic DA receptors (Kovács et al. 1985 b; Xiao et al. 1984). OXT did not affect apomorphine-induced hypomotility, a behavioral reaction thought to be a result of the stimulation of presynaptic DA receptors in the mesolimbic system (Anden et al. 1984). Since the same doses of OXT as failed to affect amphetamine-induced hyperactivity and apomorphine-induced hypoactivity were found to affect narcotic tolerance and dependence, Kovács et al. (1984 b, c) concluded that OXT is unlikely to affect narcotic addiction by interacting with presynaptic dopaminergic neuronal elements. In high doses, however, the neuropeptide blocks apomorphine-induced hypomotility in the mouse (Kovács et al. 1985 b) and rat (Xiao et al. 1984). In contrast to the previous findings, it was shown that OXT potentiates the hyperactivity induced by high doses of apomorphine, and 10^5 times lower doses of the neuropeptide are effective than on presynaptic receptors. As apomorphine-induced hyperactivity depends on the interaction of this drug with postsynaptic DA receptors, Kovács et al. (1985 b) concluded that this neuropeptide is able to change the sensitivity of postsynaptic DA receptors in the mesolimbic projection of the mouse brain. Interestingly, the dose-response effect of OXT on this behavioral measure was bell-shaped, and high doses of the neuropeptide appeared to be ineffective in mice (Kovács et al. 1985 b). Since only a high dose of OXT was employed by Xiao et al. (1984) in rats, the absence of postsynaptic interactions is in agreement with the findings of Kovács et al. (1985 b) in mice. In accordance with a postsynaptic interaction of OXT with the mesolimbic DA pathway, the neuropeptide attenuated behavioral signs of DA receptor supersensitivity following prolonged DA receptor blockade (Kovács et al. 1985 b).

Most of the data indicate that in the nigrostriatal system the situation is different and presynaptic terminals are more sensitive to OXT. In accordance with this conclusion, Schulz et al. (1979) found that the neuropeptide facilitated ipsilateral turning behavior in rats bearing unilateral lesions of the substantia nigra. It has been concluded that ipsilateral turning is a result of an increased DA output from the intact contralateral nigrostriatal terminals (Anden et al. 1966, 1984; Ungerstedt et al. 1978). Another behavioral reaction is stereotyped sniffing following high doses of apomorphine. This behavior, on the other hand, is thought to be mediated by postsynaptic DA receptors located on striatal DA-sensitive cells. In agreement with the turning data and with the conclusion of a presynaptic site of action in the nigrostriatal system, a high dose (2.5×10^{-8} mol s.c.) of OXT failed to alter apomorphine-induced sniffing in rats (Xiao et al. 1984).

In conclusion, the majority of the data indicate that OXT interacts with drug-mediated behavioral responses characteristic of both the mesolimbic DA pathway and the nigrostriatal DA projection. In the (mouse) mesolimbic system, such an interaction probably takes place at the level of DA-sensitive cells (postsynaptic effect), whereas in the (rat) nigrostriatal DA system, the neuropeptide is more likely to exert a direct effect on DA terminals or on neurons that regulate the activity of DA terminals (presynaptic effect).

6.2 Oxytocin and Brain Noradrenaline

The few data available concerning the effects of OXT on brain noradrenergic (NA-ergic) neurotransmission indicate that the neuropeptide modulates the activity of these neurons in the brain. In the hypothalamus, where NA-ergic innervation is involved in the regulation of the secretion of neurohypophyseal hormones (Swanson and Sawchenko 1983; Sladek et al. 1984; Silverman and Oldenfield 1984), the steady state level of noradrenaline (NA) is decreased by intraperitoneal (Telegdy and Kovács 1979a) or intracerebroventricular (Schwarzberg et al. 1981) OXT treatment. This decreased steady state level of NA in the hypothalamus is associated with and is a sign of increased NA utilization (Kovács and Telegdy 1983). Using a sensitive catecholamine assay, Van Heuven-Nolsen et al. (1984a) were able to show that this increased NA utilization was specifically confined to the nucleus supraopticus of the hypothalamus. In direct contrast to OXT, the intracerebroventricular injection of anti-OXT serum decreased the utilization of NA in the hypothalamus (Kovács and Telegdy 1983), suggesting that central OXT may tonically modulate the activity of hypothalamic NA terminals.

In extrahypothalamic brain structures, the action of OXT on NA-ergic neurotransmission is more controversial. In the septal area of rats, intraperitoneal OXT decreased (Telegdy and Kovács 1979a) while intracerebroventricular OXT increased the steady state level of NA (Schwarzberg et al. 1981). The utilization of NA in the septal area (measured as a whole) was unaffected by intracerebroventricular OXT treatment (Kovács and Telegdy 1983). Whereas the differences in the time course of peptide treatment may explain the difference in the steady state levels of septal NA following intraperitoneal and intracerebroventricular treatments, the failure of OXT to affect the utilization of NA in the septum may mean that different subdivisions of the septal area respond in an opposite way to intracerebroventricular OXT. In fact, Van Heuven-Nolsen et al. (1984a) have shown that low doses of intracerebroventricularly injected OXT decreased NA utilization in the medial septal nucleus, while higher doses of the neuropeptide increased NA utilization in the dorsal septal nucleus. Apart from the septal nuclei, the lateral amygdaloid nucleus was the only limbic brain site in which NA utilization was affected by intracerebroventricularly injected OXT (Van Heuven-Nolsen et al. 1984a).

In the mesencephalon, the intracerebroventricular injection of OXT failed to influence the steady state level (Schwarzberg et al. 1981) or the utilization of NA (Kovács and Telegdy 1983; Van Heuven-Nolsen et al. 1984a). Somewhat unexpectedly, however, the peripheral injection of a relatively low dose of OXT (8×10^{-10} mol s.c.) proved to be effective in increasing mesencephalic NA utilization. Whether this effect, which is observed exclusively after peripheral OXT, is related to the peripheral effects of the neuropeptide or is mediated by peripherally generated OXT fragments is a matter for speculation.

The biological significance of altered extrahypothalamic NA neurotransmission in response to OXT treatment is poorly understood. Van Heuven-Nolsen et al. (1984a) related the changes in septal NA utilization to the disturbances in fear-motivated avoidance behavior. There is a complicating factor, however, in that neurochemical lesioning of the dorsal NA-ergic bundle (the main source of input of NA-ergic fibers to the septum) only marginally influences the effects of OXT on passive avoidance behavior (Kovács et al. 1984b). Nevertheless, since the disruptive effect of OXT on passive avoidance behavior and the decreasing effect of OXT on the NA utilization in the lateral septal nucleus are consequences of acute peptide treatment, this situation is different from that where there is a lesion. Training of rats as well as later measurement of passive avoidance took place in NA-depleted rats with lesions (Kovács et al. 1984b). Therefore, these findings do not exclude the possibility that septal NA is causally involved in the effect of OXT on fear-motivated behavior in intact, nonlesioned individuals. These results suggest, however, that non-NA-ergic (c.f. section on the role of DA) effects of OXT are likewise important and that this neuropeptide is able to attenuate fear-motivated avoidance behavior even if NA-ergic input to the forebrain is lesioned (Kovács et al. 1984b).

6.3 Oxytocin and Brain Serotonin

Data relating to the interaction of OXT with brain serotonin (5-HT) are particularly few in number, in spite of the fact that brain 5-HT is involved in most behavioral processes affected by OXT (for reviews, see Essmann 1971; Kovács and Telegdy 1978, 1982). The idea that serotoninergic pathways might be important in the effects of OXT on behavioral processes is also indicated by the finding that the dorsal raphe nucleus is a brain site where local microinjection of the peptide attenuates passive avoidance behavior (Kovács et al. 1979). The raphe region, which gives rise to serotonineregic axons innervating the forebrain and the limbic and cortical regions (for a survey, see Fuxe et al. 1968), contains ten- to 20-fold higher amounts of endogenous OXT than other parts of medulla oblongata, pons or mesencephalon. Kovács and Telegdy (1982) and Schwarzberg et al. (1984) found that low quantities of intracerebroventricularly injected OXT inhibited the pargyline-induced accumulation of 5-HT in the mesencephalon. Interestingly, high doses of OXT were ineffective, suggesting an inverted U-shaped dose-response relationship for this effect of the neuropeptide as well. The same pattern of changes was observed in the 5-HT accumulation in the hippocampus (Kovács and Telegdy 1982), indicating that the raphe-hippocampal 5-HT projection might be sensitive to OXT. This is of particular interest since both the hippocampus and the raphe region are relatively rich in extrahypothalamic OXT, and local administration of the neuropeptide in these areas affects fear-motivated avoidance behavior (Kovács et al. 1979, 1980) and narcotic tolerance/dependence (Kovács et al. 1984c). Thus it can tentatively be speculated that the raphe-hippocampal 5-HT projection is involved in mediating the effects of OXT on behavior.

In the hypothalamus, the intraperitoneal injection of OXT resulted in an increase in 5-HT content (Telegdy and Kovács 1979a). However, the neuropeptide decreased the hypothalamic 5-HT upon intracerebroventricular administration (Schwarzberg et al. 1981). The pargyline-induced accumulation of hypothalamic 5-HT was dose-dependently inhibited by the intracerebroventricular injection of graded amounts of OXT (Kovács and Telegdy 1982; Schwarzberg et al. 1984). This effect, which was observed only following central OXT treatment, might be related to the endocrine action of OXT on the hypothalamus and is in line with the data of Moos and Richard (1983) suggesting that 5-HT inhibits the release of OXT from magnocellular nuclei.

6.4 Oxytocin Fragments and Cerebral Neurotransmission

The interactions of OXT fragments and analogs with brain catecholamines and indolamines have recently been reviewed by Versteeg (1983). The present section is therefore devoted to data published subsequently, to those which are critical for an understanding of peptide-transmitter interactions, and to results that have received relatively little attention in other reviews. Most OXT fragments have been produced for structure-activity studies relating to the effects of OXT on avoidance behavior or morphine addiction (for reviews, see De Wied 1976; Walter et al. 1978; De Wied and Versteeg 1979; Kovács et al. 1982a; Kovács and Telegdy 1982; Van Ree 1982; Ritzmann et al. 1984; etc.), and little or nothing is known about their occurrence, actions, and interactions in the brain. PLG is an exception in this respect, since it has been isolated from brain tissue (Nair et al. 1971) and is claimed to be the inhibiting hormone of melanocyte-stimulating hormone (Celis et al. 1971; Nair et al. 1971). The tripeptide exerts a variety of behavioral effects (see above, and see also the reviews by Barbeau et al. 1976; Kastin et al. 1975, 1980; Ehrensing et al. 1978) and has been demonstrated to interact with brain DA (Friedman et al. 1973; Versteeg et al. 1978; Szabó et al. 1981; Van Heuven-Nolsen et al. 1984b), 5-HT (Szabó et al. 1981), and opiate (Kastin et al. 1980) neurotransmission. Two further fragments that originally derived from the C-terminal part of OXT and were investigated for their interactions with brain transmission are Pro-Leu (and its protected analog Z-Pro-D-Leu) and Leu-Gly (mainly the stable analog cyclo-Leu-Gly). Both fragments inhibit the development of tolerance to and dependence on narcotic analgesics (see above). The effect of Z-Pro-D-Leu on the brain catecholamine metabolism depends on the duration of peptide treatment (Kovács et al. 1981). Following a single acute injection of the dipeptide, the steady state concentration of NA in the mouse lower brainstem (pons medulla) was decreased. Subchronic peptide treatment caused an additional decrease in brainstem DA concentration. When the turnover of catecholamines was analyzed in microdissected brain nuclei (Kovács et al. 1983b), the effect of the dipeptide revealed regional specificity. A subcutaneous dose of 10^{-7} mol, which inhibited morphine tolerance/dependence in the mouse (Kovács et al. 1981, 1984a, 1985b), increased the utilization of DA in the nucleus accumbens, a characteristic terminal projection of the mesolimbic DA pathway. This effect of the dipeptide is likely to be related to the attenuation of narcotic addiction, since abrupt changes in the activity of these DA terminals occur during narcotic addiction, and also because DA receptor blockers and synthesis inhibitors prevent the beneficial effect of the dipeptide on morphine addiction (Kovács et al. 1983c, 1984f). Although the effects of OXT and Z-Pro-D-Leu on morphine tolerance/dependence are similar, recent evidence indicates that the nonapeptide and the dipeptide exert their effects via different brain mechanisms or different receptors (Kovács et al. 1984a, 1985b).

Whereas a presynaptic interaction with DA terminals of the mesolimbic projection is suggested for Z-Pro-D-Leu (Kovács et al. 1984a, 1985b), cyclo-Leu-Gly is more likely to influence dopaminergic neurotransmission by directly influencing DA-sensitive postsynaptic receptors. In accordance with this notion, Bhargava (1984) found that cyclo-Leu-Gly prevented development of the supersensitivity of DA receptors after chronic treatment with receptor blockers. The effects of this dipeptide appear to be rather complex, since different DA pathways react oppositely to cyclo-Leu-Gly (Lee et al. 1984), and Le Douarin et al. (1984) found evidence that the mechanism by which this dipeptide inhibits DA receptor supersensitivity does not involve the direct action of the peptide on nigrostriatal dopaminergic or striatal cholinergic neurons.

In conclusion, C-terminal tri- and dipeptides derived from OXT and structural analogs of these oligopeptides are active with respect to behavioral processes and cerebral neurotransmission. The interaction of these fragments with brain DA is of functional significance for the behavioral effects.

7 Concluding Remarks

An increasing body of evidence suggests that OXT, a neuropeptide synthesized by the hypothalamic magnocellular nuclei and also present in extrahypothalamic brain regions, regulates behavioral reactions. The effects are too widespread (from learning and memory through narcotic addiction to maternal behavior) for it to be assumed that a single brain pathway or brain mechanism is specifically triggered by this neuropeptide. However, a common feature of all the known behavioral effects of OXT is that these actions are suppressive or inhibitory in nature. Under certain circumstances, such as in the case of learning/memory, this suppressive effect is obviously disadvantageous to the subject, although in selected cases of human psychopathology one might think of a therapeutic value of suppressive effects. In other situations, however, such as in the case of suppression of the adaptive response to narcotics, or in the case of the maternal effects, which can be viewed as suppression of phobia against the litter, the organism appears to benefit from the behavioral effects of OXT. Most of these behavioral effects of OXT display a feature shared by many other peptides (Kastin et al. 1984): increase of the dose of OXT does not necessarily increase its effects (e.g., on avoidance behavior, narcotic tolerance, drug-mediated responses, or transmitter metabolism), and thus bell-shaped dose-response curves are frequently observed. This in turn can result in contradictory findings, especially if a single dose of OXT is investigated. Classic cerebral neurotransmitters (e.g., DA or NA) appear to be involved in the behavioral effects of OXT, and an analysis of the mode of transmitter interactions may promote the design of behaviorally active fragments with more selective effects.

Some of the effects of OXT (e.g., on learning and memory processes or on narcotic addiction) can be produced by both peripheral and central treatments, although to different extents. Other actions (maternal behavior or behavioral arousal) are observed exclusively following central administration. Although several questions can be raised as to the interpretation of experimental studies in animals with OXT, and their relevance in human patients (Strupp et al. 1984), those effects are obviously of potential clinical interest which can also be elicited by peripheral OXT injections. All three reports on the action of OXT treatment on memory processes agree that OXT interferes with the long-term recall of memory traces (Ferrier et al. 1980; Kennett et al. 1982; Fehm-Wolsdorf et al. 1984) and are consistent with the view resulting from animal studies (Kovács and Telegdy 1982) that OXT is an amnesic neuropeptide. Further studies are required to understand the details of the role of OXT in the regulation of behavior in animals and humans.

References

Abood LG, Knapp R, Mitchell T, Booth H, Schwaab L (1980) Chemical requirements of vasopressin for barrel rotation convulsions and reversal by oxytocin. J Neurosci Res 5:191–199

Andén NE, Dahlström A, Fuxe K, Larsson K (1966) Functional role of the nigro-neostriatal dopamine neurons. Acta Pharmacol Toxicol 24:263–274

Andén NE, Alander T, Grabowska-Andén M, Liljeberg B, Lindgren S, Thornström U (1984) The pharmacology of pre- and postsynaptic dopamine receptors: differential effects of dopamine receptor agonists and antagonists. In: Usdin E, Carlsson A, Dahlström A, Engel J (eds) Catecholamines, part B: Neuropharmacology and central nervous system – theoretical aspects. Alan Liss, New York, pp 19–24

Andrews JS, Newton BA, Sahgal A (1983) The effect of vasopressin on positively rewarded responding and on locomotor activity in rats. Neuropeptides 4:17–29

Barbeau A, Roy M, Kastin AJ (1976) Double-blind evaluation of oral L-prolyl-L-leucylglycine amide in Parkinson's disease. Can Med Assoc J 114:120–122

Bhargava HN (1983) Effect of cyclo(leu-gly) on the supersensitivity of dopamine receptors in spontaneously hypertensive rats. Life Sci 32:2131–2137

Bhargava HN (1984) Enhanced striatal ^{3}H-spiroperidol binding induced by chronic haloperidol treatment inhibited by peptides administered during the withdrawal phase. Life Sci 34:873–879

Bhargava HN, Pandey RN, Matwyshyn GA (1983) Effects of prolyl-leucyl-glycinamide and cyclo(leucyl-glycine) on morphine-induced antinociception and brain μ, σ and κ opiate receptors. Life Sci 32:2095–2101

Bigl H, Stark H, Ott T, Sterba G, Matthies HJ (1977) Beeinflussung von Lernprozessen durch Hinterlappenhormone am Beispiel der Ratte. Sitzungsberichte der Akademie der Wissenschaften der DDR (Berlin, DDR) 5:84–90

Björklund A, Lindvall O, Nobin A (1975) Evidence of an incerto-hypothalamic neurone system in the rat. Brain Res 89:29–42

Bohus B, Ader B, De Wied D (1972) Effects of vasopressin on active and passive avoidance behavior. Horm Behav 3:191–197

Bohus B, Kovács GL, De Wied D (1978a) Oxytocin, vasopressin and memory: opposite effects on consolidation and retrieval processes. Brain Res 157:414–417

Bohus B, Urban I, Van Wimersma Greidanus TJB, De Wied D (1978b) Opposite effects of oxytocin and vasopressin on avoidance behavior and hippocampal theta rhythm in the rat. Neuropharmacology 17:239–247

Bohus B, Conti L, Kovács GL, Versteeg DHG (1982) Modulation of memory processes by neuropeptides: interaction with neurotransmitter systems. In: Marsan CA, Matthies HJ (eds) Neuronal plasticity and memory formation. Raven, New York, pp 75–87

Brinton RE, Wamsley JK, Gee KW, Wan YP, Yamamura HI (1984) (^{3}H)Oxytocin binding sites in the rat brain demonstrated by quantitative light microscopic autoradiography. Eur J Pharmacol 102:365–367

Buijs RM (1978) Intra- and extrahypothalamic vasopressin and oxytocin pathways in the rat. Pathways to the limbic system, medulla oblongata and spinal cord. Cell Tissue Res 192:423–435
Buijs RM (1983) Vasopressin and oxytocin – their role in neurotransmission. Pharmacol Ther 22:127–141
Buijs RM, Van Heerikhuize JJ (1982) Vasopressin and oxytocin release in the brain: a synaptic event. Brain Res 252:71–76
Burbach JPH, De Wied D (1981) Memory effects and brain proteolysis of neurohypophyseal hormones. In: Schlesinger DH (ed) Neurohypophyseal peptide hormones and other biologically active peptides. Elsevier/North Holland, Amsterdam, pp 69–87
Burbach JPH, Lebouille JLM (1983) Proteolytic conversion of arginine-vasopressin and oxytocin by brain synaptic membranes. J Biol Chem 258:1487–1494
Burbach JPH, Bohus B, Kovács GL, Van Nispen JW, Greven HM, De Wied D (1983 a) Oxytocin is a precursor of potent behaviourally active neuropeptides. Eur J Pharmacol 94:125–131
Burbach JPH, Kovács GL, Wang XC, De Wied D (1983 b) Metabolites of arginine-vasopressin and oxytocin are highly potent neuropeptides in the brain. In: Koch G, Richter D (eds) Biochemical and clinical aspects of neuropeptides: synthesis, processing and gene structure. Academic, Orlando, pp 211–222
Caldwell JD, Pedersen CA, Prange AJ (1984) Oxytocin facilitates sexual behavior in estrogen-treated ovariectomized rats. J Steroid Biochem 20:1510
Celis ME, Taleisnik S, Walter R (1971) Regulation of formation and proposed structure of the factor inhibiting the release of melanocyte-stimulating hormone. Proc Natl Acad Sci (USA) 68:1428–1433
Chard T, Hudson CN, Edwards CRW, Boyd NRH (1971) Release of oxytocin and vasopressin by the human fetus during labour. Nature 234:352–354
Cohn ML, Cohn M (1975) Barrel rotation induced by somatostatin in the non-lesioned rat. Brain Res 96:138–141
Contreras P, Takemori AE (1981) Facilitation of morphine-induced tolerance and physical dependence by prolyl-leucyl-glycinamide. Eur J Pharmacol 71:259–268
Cools AR, Broekkamp CLE, Gieles LCM, Megens A, Mortieaux HJGM (1977) Site of action of development of partial tolerance to morphine in cats. Psychoneuroendocrinology 2:17–33
Costall B, Domeney AM, Naylor RJ (1984) Long-term consequences of antagonism by neuroleptics of behavioural events occurring during mesolimbic dopamine infusion. Neuropharmacology 23:287–294
De Kloet ER, Rotteveel F, Voorhuis TD, Terlou M (1985) Topography of binding sites for neurohypophyseal hormones in rat brain. Eur J Pharmacol 110:113–119
Delanoy RL, Dunn AJ, Tintner R (1978) Behavioral responses to intracerebroventricularly administered neurohypophyseal peptides in mice. Horm Behav 11:348–362
Destrade C, Jafford R (1978) Post-trial hippocampal and lateral hypothalamic electrical stimulation. Facilitation of long-term memory of appetititive and avoidance learning tasks. Behav Biol 22:354–374
De Wied D (1965) The influence of posterior and intermediate lobe of the pituitary and pituitary peptides on the maintenance of conditioned avoidance response in rats. Int J Neuropharmacol 4:157–167
De Wied D (1969) Effects of peptide hormones on behavior. In: Ganong W, Martini L (eds) Frontiers in neuroendocrinology. Oxford University Press, New York, pp 97–140
De Wied D (1976) Behavioral effects of intraventricularly administered vasopressin and vasopressin fragments. Life Sci 19:685–690
De Wied D (1979) Pituitary neuropeptides and behavior. In: Fuxe K, Hökfelt T, Luft R (eds) Central regulation of the endocrine system. Plenum, New York, pp 297–314
De Wied D, Bohus B (1966) Long term and short term effect on retention of a conditioned avoidance response in rats by treatment respectively with long acting pitressin or α-MSH. Nature 212:1484–1486
De Wied D, Bohus B (1979) Modulation of memory processes by neuropeptides of hypothalamic-neurohypophyseal origin. In: Brazier MAB (ed) Brain mechanisms in memory and learning: from the single neuron to man. Raven, New York, pp 139–149

De Wied D, Gispen WH (1976) Impaired development of tolerance to morphine analgesia in rats with hereditary diabetes insipidus. Psychopharmacology 46:27–29

De Wied D, Gispen WH (1977) Behavioral effects of peptides. In: Gainer H (ed) Peptides in neurobiology. Plenum, New York, pp 397–448

De Wied D, Versteeg DHG (1979) Neurohypophyseal principles and memory. Fed Proc 38:2348–2354

De Wied D, Van Wimersma Greidanus TJB, Bohus B, Urban I, Gispen WH (1976) Vasopressin and memory consolidation. In: Corner MA, Swaab DF (eds) Perspectives in brain research. Prog Brain Res 45:181–191

De Wied D, Gaffori O, Van Ree JM, De Jong W (1984) Central target for the behavioural effects of vasopressin neuropeptides. Nature 308:276–278

Dickinson SL, Slater P (1980) Opiate receptor antagonism by L-prolyl-L-leucyl-glycinamide, MIF-I. Peptides 1:293–299

Dogterom J, Snijdewindt FGM, Buijs RM (1978) The distribution of vasopressin and oxytocin in the rat brain. Neurosci Lett 9:341–346

Drago F, Bohus B, De Wied D (1981) Interaction between vasopressin and oxytocin in the modulation of passive avoidance retention of the rat. Neurosci Lett [Suppl] 7:S260

Drago F, Kovács GL, Scapagnini U (1984) Prolactin-induced behavioral effects and opioids. In: Delitalia G, Motta M, Serio M (eds) Opioid modulation of endocrine function. Frontiers in neuroscience. Raven, New York, pp 137–145

Drago F, Kovács GL, Szabó G, Scapagnini U, Telegdy G (1985) Effects of haloperidol on morphine-induced analgesia, morphine tolerance and withdrawal in hyperprolactinaemic rats. Neuropharmacology 24:1027–1031

Ehrensing RH, Kastin AJ, Larsons PF, Bishop GA (1978) Melanocyte-stimulating-hormone-release-inhibiting factor-I and tardive dyskinesia. Dis Nerv Syst 38:303–307

Esposito R, Kornetsky C (1977) Morphine lowering of self-stimulation thresholds: lack of tolerance with long-term administration. Science 195:189–191

Essmann WB (1971) The role of biogenic amines in memory consolidation. In: Adám G (ed) Biology of memory. Akadémiai Kiadó/Plenum, Budapest, pp 213–238

Ettenberg A, Van der Koy D, Le Moal M, Koob GF, Bloom FE (1983) Can aversive properties of (peripherally-injected) vasopressin account for its putative role in memory? Behav Brain Res 7:331–350

Fahrbach SE, Morrell JI, Pfaff DW (1985) Role of oxytocin in the onset of estrogen-facilitated maternal behavior. In: Amico JA, Robinson AG (eds) Oxytocin: clinical and laboratory aspects. Elsevier, Amsterdam, pp 372–388

Fehm-Wolsdorf G, Born J, Voigt KH, Fehm HL (1984) Human memory and neurohypophyseal hormones: opposite effects of vasopressin and oxytocin. Psychoneuroendocrinology 9:285–292

Ferrier BM, Kennett DJ, Devlin MC (1980) Influence of oxytocin on human memory processes. Life Sci 27:2311–2317

Ferrier BM, McClorry SA, Cochrane AW (1983) Specific binding of (^{3}H)oxytocin in female rat brain. Can J Physiol Pharmacol 61:989–995

Feuerstein G, Zerbe RL, Faden I (1984) Central cardiovascular effects of vasotocin, oxytocin and vasopressin in conscious rats. J Pharmacol Exp Ther 228:348–353

Folley SJ, Knaggs GS (1966) Milk-ejection activity (oxytocin) in the external jugular vein blood of the cow, goat and sow, in relation to the stimulus of licking and suckling. J Endocrinol 34:197–214

Friedman E, Friedman J, Gershon S (1973) Dopamine synthesis: stimulation by a hypothalamic factor. Science 182:831–832

Fuxe K, Hökfelt T, Ungerstedt U (1968) Localization of indolalkylamines in the CNS. In: Garattini S, Shore PE (eds) Advances in pharmacology, vol 6, part A, Academic, New York, pp 235–251

Gash DM, Thomas GJ (1983) What is the importance of vasopressin in memory processes? Trends Neurosci 60:197–198

Gibbs DM (1984) Dissociation of oxytocin, vasopressin and corticotropin secretion during different types of stress. Life Sci 35:487–491

Hagan JJ, Bohus B (1984) Vasopressin prolongs bradycardiac response during orientation. Behav Neural Biol 41:77–83

Hawthorn J, Ang VTY, Jenkins JS (1984) Comparison of the distribution of oxytocin and vasopressin in the rat brain. Brain Res 307:289–294
Himmelsbach CK (1943) With reference to physical dependence. Fed Proc 2:201–203
Hoffmann PL, Ritzmann RF, Walter R, Tabakoff B (1978) Arginine vasopressin maintains ethanol tolerance. Nature 276:614–616
Izquierdo I, Perry ML, Dias RD, Souza DO, Elisabetsky E, Carrasco MA, Orsingher OA, Netto CA (1981) Endogenous opioids, memory modulation, and state dependency. In: Martinez JL, Jensen RA, Messing RB, Rigter H, McGaugh JL (eds) Endogenous peptides and learning and memory processes. Academic, New York, pp 269–290
Joëls M, Urban IJA (1982) The effect of microiontophoretically applied vasopressin and oxytocin on single neurones in the septum and dorsal hippocampus of the rat. Neurosci Lett 33:79–84
Judge MA, Quartermain D (1982) Characteristics of retrograde amnesia following reactivation of memory in mice. Physiol Behav 28:585–590
Kastin AJ, Plotnikoff NP, Sandman CA, Spirtes MA, Kostrzewa RM, Paul SM, Stratton LO, Miller LH, Labrie F, Schally AV, Goldman H (1975) The effects of MSH and MIF on the brain. In: Stumpf WE, Grant LD (eds) Anatomical neuroendocrinology. Karger, Basel, pp 290–297
Kastin AJ, Nissen C, Zadina JE, Schally AV, Ehrensing RH (1980) Naloxone-like actions of MIF-1 do not require the presence of the pituitary. Pharmacol Biochem Behav 13:907–912
Kastin AJ, Zadina JE, Banks WA, Graf MV (1984) Misleading concepts in the field of brain peptides. Peptides 5:249–253
Kennett DJ, Devlin MC, Ferrier BM (1982) Influence of oxytocin on human memory processes: validation by a control study. Life Sci 31:273–275
Koob GF, Le Moal M, Gaffori O, Manning M, Sawyer WH, Rivier J, Bloom FE (1981) Arginine vasopressin and a vasopressin antagonist peptide: opposite effects on extinction of active avoidance in rats. Regul Peptides 2:153–163
Kordower JH, Bodnar RJ (1984) Vasopressin analgesia: specificity of action and non-opioid effects. Peptides 5:747–756
Kovács GL, De Wied D (1981) Endorphin influences on learning and memory. In: Martinez JL, Jensen RA, Messing RB, Rigter H, McGaugh JL (eds) Endogenous peptides and learning and memory processes. Academic, New York, pp 231–247
Kovács GL, De Wied D (1983) Hormonally active arginine-vasopressin suppresses endotoxin-induced fever in rats: lack of effect of oxytocin and a behaviorally active vasopressin fragment. Neuroendocrinology 37:258–261
Kovács GL, Telegdy G (1978) Indoleamines and behavior. The possible role of serotoninergic mechanisms in the pituitary-adrenocortical hormone-induced behavioral actions. In: Lissák K (ed) Recent progress of neurobiology in Hungary, vol 7. Akadémiai Kiadó, Budapest, pp 31–97
Kovács GL, Telegdy G (1982) Role of oxytocin in memory and amnesia. Pharmacol Ther 18:375–395
Kovács GL, Telegdy G (1983) Effects of oxytocin, des-glycinamide-oxytocin and anti-oxytocin serum on the α-MPT-induced disappearance of catecholamines in the rat brain. Brain Res 268:307–314
Kovács GL, Telegdy G (1984) Oxytocin diminishes narcotic addiction: effects on morphine tolerance/withdrawal and heroin self-administration. Neurosci Lett [Suppl] 18:S354
Kovács GL, Telegdy G (1985) Oxytocin in memory and reinforcement. In: Amico J, Robinson AG (eds) Oxytocin: clinical and laboratory studies. Elsevier, Amsterdam, pp 359–371
Kovács GL, Vécsei L, Sazbó G, Telegdy G (1977) The involvement of catecholaminergic mechanisms in the behavioural action of vasopressin. Neurosci Lett 5:337–344
Kovács GL, Vécsei L, Telegdy G (1978) Opposite action of oxytocin to vasopressin in passive avoidance behavior in rats. Physiol Behav 20:801–802
Kovács GL, Bohus B, Versteeg DHG, De Kloet ER, De Wied D (1979) Effect of oxytocin and vasopressin on memory consolidation: sites of action and catecholaminergic correlates after local microinjection into limbic-midbrain structures. Brain Res 175:303–314

Kovács GL, Bohus B, Versteeg DHG (1980) The interaction of posterior pituitary neuropeptides with monoaminergic neurotransmission: significance in learning and memory processes. Prog Brain Res 53:123–140

Kovács GL, Szontágh L, Baláspiri L, Hódi K, Bohus P, Telegdy G (1981) On the mode of action of an oxytocin derivative (Z-Pro-D-Leu) on morphine dependence in mice. Neuropharmacology 20:647–651

Kovács GL, Bohus B, Versteeg DHG, Telegdy G, De Wied D (1982a) Neurohypophyseal hormones and memory. In: Yoshida H, Hagihara Y, Ebashi S (eds) Advances in pharmacology and therapeutics II, vol 1. CNS pharmacology. Neuropeptides, Pergamon, Oxford, pp 175–187

Kovács GL, Buijs RM, Bohus B, Van Wimersma Greidanus TJB (1982b) Microinjection of arginine^8-vasopressin antiserum into the dorsal hippocampus attenuates passive avoidance behavior in rats. Physiol Behav 28:45–48

Kovács GL, Ribárszki Z, Telegdy G (1983a) Reversal of electroconvulsive shock-induced amnesia by neuropeptides. In: Endröczi E, De Wied D, Angelucci L, Scapagnini U (eds) Neuropeptides and psychosomatic processes. Akadémiai Kiadó, Budapest, pp 167–174

Kovács GL, Acsai L, Tihanyi A, Faludi M, Telegdy G (1983b) Influence of Z-prolyl-D-leucine on α-MPT-induced catecholamine utilization in specific mouse brain nuclei. Pharmacol Biochem Behav 18:345–349

Kovács GL, Acsai L, Tihanyi A, Telegdy G (1983c) Catecholamine utilization in distinct mouse brain nuclei during acute morphine treatment, morphine tolerance and withdrawal syndrome. Eur J Pharmacol 93:149–158

Kovács GL, Izbéki F, Horváth Z, Telegdy G (1984a) Effects of oxytocin and a derivative (Z-prolyl-D-leucine) on morphine tolerance/dependence are mediated by the limbic system. Behav Brain Res 14:1–8

Kovács GL, Schwarzberg H, Veldhuis HD, Telegdy G (1984b) Influences of oxytocin on behavioral processes: role of monoaminergic neurotransmission. Acta Physiol Hung 63:249

Kovács GL, Horváth Z, Falkay G (1984c) Brain dopamine and neuropeptide interactions in narcotic addiction: I. The role of pre- and postsynaptic receptors. Annual meeting of the Hungarian physiological society, Szeged, July 5–7, 1984, p 55

Kovács GL, Telegdy G, Hódi K (1984d) Drugs affecting brain dopamine interfere with the effect of Z-prolyl-D-leucine on morphine withdrawal. Pharmacol Biochem Behav 21:345–348

Kovács GL, Borthaiser Z, Telegdy G (1985a) Oxytocin reduces heroin self-administration in heroin-tolerant rats. Life Sci 37:17–26

Kovács GL, Horváth Z, Sarnyai Z, Faludi M, Telegdy G (1985b) Oxytocin and a C-terminal derivative (Z-prolyl-D-leucine) attenuate tolerance to and dependence on morphine and interact with dopaminergic neurotransmission in the mouse brain. Neuropharmacology 24:413–419

Kovács GL, Telegdy G, Laczi F, László F (1985c) Oxytocin and vasopressin in memory and amnesia. In: Will BE, Schmitt P, Dalrymple-Alford JC (eds) Brain plasticity, learning, and memory. Plenum, New York, pp 297–301

Kovács GL, Vecsernyés M, Laczi F, Faludi M, Telegdy G, László FA (1985d) Acute morphine treatment and morphine tolerance/dependence alter immunoreactive oxytocin levels in the mouse hippocampus. Brain Res 328:158–160

Kozlowski GP, Nilaver G, Zimmermann EA (1983) Distribution of neurohypophyseal hormones in the brain. Pharmacol Ther 21:325–349

Krivoy WA, Zimmermann E, Lande S (1974) Facilitation of development of resistance to morphine analgesia by desglycinamide^9-lysine vasopressin. Proc Natl Acad Sci (USA) 71:1852–1856

Kruse H, Van Wimersma Greidanus TJB, De Wied D (1977) Barrel rotation by vasopressin and related peptides in rats. Pharmacol Biochem Behav 7:311–313

Landfield PW, McGaugh JL (1972) Effects of electroconvulsive shock and brain stimulation on EEG cortical theta rhythms in rats. Behav Biol 7:271–278

Landfield PW, McGaugh JL, Tusa RJ (1972) Theta rhythm: a temporal correlate of memory storage processes in the rat. Science 175:87–89

Lee JM, Ritzmann RF, Fields JZ (1984) Cyclo(leu-gly) has opposite effects on D-2 dopamine receptors in different brain areas. Peptides 5:7–10
Le Douarin C, Fage D, Scatton B (1984) Effects of cyclo(leu-gly) on neurochemical indices of striatal dopaminergic supersensitivity induced by prolonged haloperidol treatment. Life Sci 34:393–399
Le Piane FG, Phillips AG (1978) Differential effects of electrical stimulation of the amygdala, caudate-putamen or substantia nigra pars compacta on taste aversion and passive avoidance in rats. Physiol Behav 21:979–985
Maroli AN, Tsang WK, Stutz RM (1978) Morphine and self-stimulation: evidence for action on a common neural substrate. Pharmacol Biochem Behav 8:119–123
McGaugh JL (1966) Time-dependent processes in memory storage. Science 153:1351–1358
McGaugh JL (1973) Drug facilitation of learning and memory. Ann Rev Pharmacol 13:229–241
McGaugh JL, Dawson RG (1971) Modification of memory storage processes. In: Honig WK, James PHR (eds) Animal memory. Academic, New York, pp 215–242
McGaugh JL, Gold PE, Handwerker MJ, Jensen RA, Martinez JL, Meligeni JA, Vasquez BJ (1979) Altering memory by electrical and chemical stimulation of the brain. In: Brazier MAB (ed) Brain mechanisms in memory and learning: from single neuron to man. IBRO Monograph Series, vol 4. Raven, New York, pp 151–164
Meisenberg G (1981) Short-term behavioral effects of posterior pituitary peptides in mice. Peptides 2:1–8
Meisenberg G (1982) Short-term behavioural effects of neurohypophyseal hormones: pharmacological characteristics. Neuropharmacology 21:309–316
Meisenberg G, Simmons WH (1982) Behavioral effects of intracerebroventricularly administered neurohypophyseal hormone analogs in mice. Pharmacol Biochem Behav 16:819–825
Meisenberg G, Simmons WH (1984 a) Amastatin potentiates the behavioral effects of vasopressin and oxytocin in mice. Peptides 5:535–539
Meisenberg G, Simmons WH (1984 b) Factors involved in the inactivation of vasopressin after intracerebroventricular injection in mice. Life Sci 34:1231–1240
Mens WBJ, Van Egmond MA, De Rotte AA, Van Wimersma Greidanus TJB (1982) Neurohypophyseal peptide levels in CSF and plasma during passive avoidance behavior in rats. Horm Behav 16:371–382
Mens WBJ, Witter A, Van Wimersma Greidanus TJB (1983) Penetration of neurohypophyseal hormones from plasma into cerebrospinal fluid (CSF): half-times of disappearance of these peptides from CSF. Brain Res 262:143–149
Moos F, Richard P (1983) Serotonergic control of oxytocin release during suckling in the rat: opposite effects in conscious and anesthetized rats. Neuroendocrinology 36:300–306
Nair RMG, Kastin AJ, Schally AV (1971) Isolation and structure of hypothalamic MSH release-inhibiting hormone. Biochem Biophys Res Commun 43:1376–1381
Obrist P, Webb RA, Sutterer JR, Howard JL (1970) The cardiac-somatic relationship: some reformulations. Psychophysiology 6:569–587
Palkovits M, Browstein M (1983) Extrahypothalamic distribution and action of hypothalamic hormones. In: Iversen LL, Iversen SD, Snyder SH (eds) Handbook of psychopharmacology, vol 16, Neuropeptides, Plenum, New York, pp 467–487
Pedersen CA, Prange AJ (1979) Induction of maternal behavior in virgin rats after intracerebroventricular administration of oxytocin. Proc Natl Acad Sci (USA) 76:6661–6665
Pedersen CA, Ascher JA, Monroe YL, Prange AJ (1982) Oxytocin induces maternal behavior in virgin female rats. Science 216:648–650
Pedersen CA, Caldwell JD, Prange AJ (1984) Oxytocin antiserum inhibits the onset of ovarian steroid-induced maternal behavior. Symposium on Oxytocin, Lac Beauport, Canada
Porsolt RD, Bertin A, Jalfre M (1977) Behavioural despair in mice: a primary screening test for antidepressants. Arch Int Pharmacodyn Ther 229:327–336
Quartermain D, Freedman LS, Botwinick CY, Gutwein BM (1977) Reversal of cycloheximide-induced amnesia by adrenergic receptor stimulation. Pharmacol Biochem Behav 7:259–267

Reppert SM, Perlow MJ, Artman HG, Ungerleider LG; Fisher DA, Klein DC (1984) The circadian rhythm of oxytocin in primate cerebrospinal fluid: effects of destruction of the suprachiasmatic nuclei. Brain Res 307:384–387

Rigter H, Van Riezen H (1978) Hormones and memory. In: Lipton MA, De Mascio A, Killam AF (eds) Psychopharmacology: a generation of progress. Raven, New York, pp 677–689

Rigter H, Van Riezen H, De Wied D (1974) The effects of ACTH- and vasopressin-analogues on CO_2-induced retrograde amnesia in rats. Physiol Behav 13:381–388

Rigter H, Rijk H, Crabbe JC (1980) Tolerance to ethanol and the severity of withdrawal in mice are enhanced by a vasopressin fragment. Eur J Pharmacol 64:53–68

Ritzmann RF, Walter R, Bhargava HN, Krivoy W (1980) The inhibition of the development of tolerance to and dependence on morphine by peptides. In: Ajmone Marsan C, Traczyk WZ (eds) Neuropeptides and neural transmission. Raven, New York, pp 237–244

Ritzmann RF, Colbern DL, Zimmermann EG, Krivoy W (1984) Neurohypophyseal hormones in tolerance and physical dependence. Pharmacol Ther 23:281–312

Robinson AG, Zimmermann EA (1973) Cerebrospinal fluid and ependymal neurophysin. J Clin Invest 52:1260–1267

Robinson ICAF, Jones PM (1982) Neurohypophyseal peptides in cerebrospinal fluid: recent studies. In: Baertschi AJ, Dreifuss JJ (eds) Neuroendocrinology of vasopressin, corticoliberin and opiomelanocortins. Academic, London, pp 21–31

Rubin BS, Menniti FS, Bridges RS (1983) Intracerebroventricular administration of oxytocin and maternal behavior in rats after prolonged and acute steroid pretreatment. Horm Behav 17:45–53

Sahgal A, Wright C (1984) Choice, as opposed to latency, measures in avoidance suggest that vasopressin and oxytocin do not affect memory in rats. Neurosci Lett 48:299–304

Sahgal A, Keith AB, Wright C, Edwardson JA (1982) Failure of vasopressin to enhance memory in a passive avoidance task in rats. Neurosci Lett 28:87–92

Sara SJ, Barnett J, Toussaint P (1982) Vasopressin accelerates appetitive discrimination learning and impairs its reversal. Behav Proc 7:157–167

Schmidt WK, Holaday JW, Loh HH, Way EL (1978) Failure of vasopressin and oxytocin to antagonize acute morphine antinociception or facilitate narcotic tolerance development. Life Sci 23:151–158

Schulz H, Kovács GL, Telegdy G (1974) Effect of physiological doses of vasopressin and oxytocin on avoidance and exploratory behaviour in rats. Acta Physiol Hung 45:211–215

Schulz H, Kovács GL, Telegdy G (1976) The effect of vasopressin and oxytocin on avoidance behavior in rats. In: Endröczi E (ed) Cellular and molecular bases of neuroendocrine processes. Akadémiai Kiadó, Budapest, pp 555–564

Schulz H, Kovács GL, Telegdy G (1979) Action of posterior pituitary neuropeptides on the nigro-striatal dopaminergic system. Eur J Pharmacol 57:185–190

Schwarzberg H, Unger H (1970) Änderung der Reaktionszeit von Ratten nach Applikation von Vasopressin, Oxytocin und Na-thioglykolat. Acta Biol Med Germ 24:507–516

Schwarzberg H, Hartmann G, Kovács GL, Telegdy G (1976) The effect of intraventricular administration of oxytocin and vasopressin on self-stimulation in rats. Acta Physiol Hung 47:127–131

Schwarzberg H, Betschen K, Unger H, Schulz H (1978) Beeinflussung der hypothalamischen Selbststimulation durch intrazerebroventriculär verabreichtes Vasopressin und Oxytocin. Acta Biol Med Germ 37:1295–1296

Schwarzberg H, Kovács GL, Szabó G, Telegdy G (1981) Intraventricular administration of vasopressin and oxytocin affects the steady-state levels of serotonin, dopamine and norepinephrine in rat brain. Endocrinol Exp (Bratisl) 15:75–80

Schwarzberg H, Kovács GL, Telegdy G (1984) The influence of oxytocin on the steady-state level and accumulation of serotonin in rat brain regions. Neuropeptides 4:145–156

Siegel S (1975) Evidence from rats that morphine tolerance is a learned response. J Comp Physiol Psychol 89:498–506

Siegel S (1976) Morphine analgesic tolerance: its situation specificity supports a Pavlovian conditioning model. Science 193:323–325
Siegel S (1978) Tolerance to the hypothermic effect of morphine in the rat is a learned response. J Comp Physiol Psychol 92:1137–1149
Silverman AJ, Oldenfield BJ (1984) Synaptic input to vasopressin neurons of the paraventricular nucleus (PVN). Peptides 5:139–150
Singhal RL, Rastogi RB (1982) MIF-1: effects on norepinephrine, dopamine and serotonin metabolism in certain discrete brain regions. Pharmacol Biochem Behav 16:229–233
Sladek JR, Fields J, Phelps CJ, Khachaturian H (1984) Development of the catecholamine innervation of the supraoptic nucleus in the Brattleboro rat. Peptides 5 [Suppl 1]:151–155
Sofroniew MV (1983) Vasopressin and oxytocin in the mammalian brain and spinal cord. Trends Neurosci 6:467–472
Sterba G (1974) Ascending neurosecretory pathways of the peptidergic type. In: Knowles F, Vollrath L (eds) Neurosecretion – the final neurosecretory pathway. Springer, Berlin Heidelberg New York, pp 38–47
Sterba G, Naumann W, Hoheisel G (1980) Exohypothalamic axons of the classical neurosecretory system and their synapses. In: McConnell PS, Boer GJ, Romijn HJ, Van de Poll NE, Corner MA (eds) Adaptive capabilities of the nervous system. Prog Brain Res 53:141–158
Strupp B, Weingartner H, Goodwin FK, Gold PW (1984) Neurohypophyseal hormones and cognition. Pharmacol Ther 23:267–279
Swanson HH, Bolwerk E (1984) Does oxytocin play a role in the onset of maternal behaviour? J Steroid Biochem 20:1510
Swanson LW, Sawchenko PE (1983) Hypothalamic integration: organization of the paraventricular and supraoptic nuclei. Ann Rev Neurosci 6:269–324
Szabó G, Kovács GL, Baláspiri L, Telegdy G (1981) Dose-related effect of the oxytocin fragment prolyl-leucyl-glycinamide on α-MPT-induced catecholamine disappearance and serotonin levels in rat brain. Neurochem Int 3:411–416
Szabó G, Kovács GL, Telegdy G (1983) The effect of oxytocin and an oxytocin fragment (prolyl-leucyl-glycinamide) on the development of ethanol tolerance. Acta Endocrinol 103 [Suppl 256]:242
Szabó G, Kovács GL, Telegdy G (1984) The effect of oxytocin and two of its fragments on ethanol tolerance in mice. J Steroid Biochem 20:1512
Szabó G, Kovács GL, Székeli S, Baláspiri L, Telegdy G (1985) C-terminal fragments of oxytocin (prolyl-leucyl-glycinamide and Z-prolyl-D-leucine) attenuate the development of tolerance to ethanol. Acta Physiol Hung (in press)
Székely JI, Miglécz E, Dunai-Kovács Z, Tarnawa I, Rónai AZ, Gráf L, Bajusz S (1979) Attenuation of morphine tolerance and dependence by α-melanocyte-stimulating hormone (α-MSH). Life Sci 24:1931–1938
Tabakoff B, Ritzmann RF (1977) The effect of 6-hydroxydopamine on tolerance to and dependence on ethanol. J Pharmacol Exp Ther 203:319–321
Tanaka M, Versteeg DHG, De Wied D (1977) Regional effects of vasopressin on rat brain catecholamine metabolism. Neurosci Lett 4:321–325
Telegdy G, Kovács GL (1979a) Role of monoamines in mediating the action of hormones on learning and memory. In: Brazier MAB (ed) Brain mechanisms in memory and learning: from single neuron to man. IBRO Monograph Series, vol 4. Raven, New York, pp 249–268
Telegdy G, Kovács GL (1979b) Role of monoamines in mediating the action of ACTH; vasopressin and oxytocin. In: Collu R, Barbeau A, Ducharme JR, Rochefort JG (eds) Central nervous system effects of hypothalamic hormones and other peptides. Raven, New York, pp 189–205
Thompson T, Pickens R (1975) An experimental analysis of behavioral factors in drug dependence. Fed Proc 34:1759–1770
Tindal JS (1974) Stimuli that cause the release of oxytocin. In: Knobil E, Sawyer WH (eds) The pituitary gland and its neuroendocrine control, part 1. American Physiological Society, Washington, pp 257–267 (Handbook of physiology, vol 4, sect 7: Endocrinology)

Unger H (1977) Funktionelle Aspekte der Informationsübermittelung durch die Oligopeptide Vasopressin und Oxytocin bei Säugetieren. Sitzungsberichte der Akademie der Wissenschaften der DDR, Berlin, DDR, 5:62–83

Ungerstedt U, Ljungberg T, Schultz W (1978) Dopamine receptor mechanisms: behavioral and electrophysiological studies. In: Roberts JL, Woodruff GN, Iversen LL (eds) Advances in biochemical pharmacology, vol 19. Raven, New York, pp 311–321

Urban IJA (1981) Intraseptal administration of vasopressin and oxytocin affects hippocampal electroencephalogram in rat. Exp Neurol 73:131–147

Van Heuven-Nolsen D, Versteeg DHG (1985) Interaction of vasopressin with the nigrostriatal dopamine system: site and mechanism of action. Brain Res 337:269–276

Van Heuven-Nolsen D, De Kloet ER, Versteeg DHG (1984a) Oxytocin affects noradrenaline utilization in distinct limbic-forebrain regions of the rat brain. Neuropharmacology 23:269–276

Van Heuven-Nolsen D, De Kloet ER, Versteeg DHG (1984b) Pro-Leu-GlyNH$_2$ affects dopamine and noradrenaline utilization in rat limbic-forebrain nuclei. Brain Res 322:213–218

Van Ree JM (1982) Neurohypophyseal hormones and addiction. In: Yoshida H, Hagihara Y, Ebashi S (1982) Advances in pharmacology and therpeutics II, vol 1. CNS pharmacology. Neuropeptides. Pergamon, Oxford, pp 199–209

Van Ree JM (1983) Neuropeptides and addictive behaviour. Alcohol Alcoholism 18:325–330

Van Ree JM, De Wied D (1976) Prolyl-leucyl-glycinamide (PLG) facilitates morphine dependence. Life Sci 19:1331–1339

Van Ree JM, De Wied D (1977a) Modulation of heroin self-administration by neurohypophyseal principles. Eur J Pharmacol 43:199–202

Van Ree JM, De Wied D (1977b) Heroin self-administration is under control of vasopressin. Life Sci 21:315–320

Van Wimersma Greidanus TJB, Dogterom J, De Wied D (1975) Intraventricular administration of anti-vasopressin serum inhibits memory consolidation of rats. Life Sci 16:637–644

Van Wimersma Greidanus TJB, Van Ree JM, Versteeg DHG (1980) Neurohypophyseal peptides and avoidance behavior: the involvement of vasopressin and oxytocin in memory processes. In: Ajmone Marsan C, Traczyk WZ (eds) Neuropeptides and neural transmission. Raven, New York, pp 293–300

Van Wimersma Greidanus TJB, Bohus B, Kovács GL, Versteeg DHG, Burbach JHP, De Wied D (1983) Sites of behavioral and neurochemical action of ACTH-like peptides and neurohypophyseal hormones. Neurosci Biobehav Rev 7:453–463

Versteeg DHG (1983) Neurohypophyseal hormones and brain neurochemistry. Pharmacol Ther 19:297–325

Versteeg DHG, Tanaka M, De Kloet ER, Van Ree JM, De Wied D (1978) Prolyl-leucyl-glycinamide (PLG): regional effects on α-MPT-induced catecholamine disappearance in rat brain. Brain Res 143:561–566

Walter R, Hoffmann PL, Flexner JB, Flexner LB (1975) Neurohypophyseal hormones, analogs and fragments: their effect on puromycin-induced amnesia. Proc Natl Acad Sci (USA) 72:4180–4184

Walter R, Van Ree JM, De Wied D (1978) Modification of conditioned behavior of rats by neurohypophyseal hormones and analogues. Proc Natl Acad Sci (USA) 75:2493–2496

Way EL, Rezvani A (1984) Opiate tolerance and physical dependence: assessment and mechanisms. In: Hughes J, Collier HOJ, Rance MJ, Tyers MB (eds) Opioids past, present and future. Taylor and Francis, London, pp 103–108

Wikler A (1980) Opioid dependence. Mechanism and treatment. Plenum, New York

Xiao X, Veldhuis HD, Van Ree JM (1984) Neuropeptides related to neurohypophyseal hormones interfere with apomorphine-induced behavioral changes. Neuropeptides 4:237–245

Oxytocin as an Ovarian Hormone

D. C. WATHES, R. W. SWANN, D. G. PORTER and B. T. PICKERING

Contents

1 Introduction

The synthesis, secretion and actions of hypothalamic oxytocin are covered extensively in the rest of this volume. However, a number of recent reports have confirmed an earlier observation made by Ott and Scott (1910) that the corpus luteum also contains a hormone with an oxytocic action. We shall examine the evidence that this luteal hormone is authentic oxytocin which is synthesized within the ovary. The function of luteal oxytocin is still not clear, but various suggestions have been made and the evidence to support these will be examined. Immunoreactive oxytocin has since been found in the testis and a variety of other glands and these will also be discussed.

2 Characterization of Luteal Oxytocin

Luteal oxytocin was first identified on the basis of its biological activity. It was found that acidic extracts of non-pregnant ovine corpus luteum caused a significant increase in resting tension and stimulated contractions in a rat uterine strip bioassay (Wathes and Swann 1982; Wathes et al. 1983b). An extract of bovine

Department of Anatomy, The Medical School, University of Bristol, Bristol BS8 1TD, Great Britain

Current Topics in Neuroendocrinology, Vol. 6

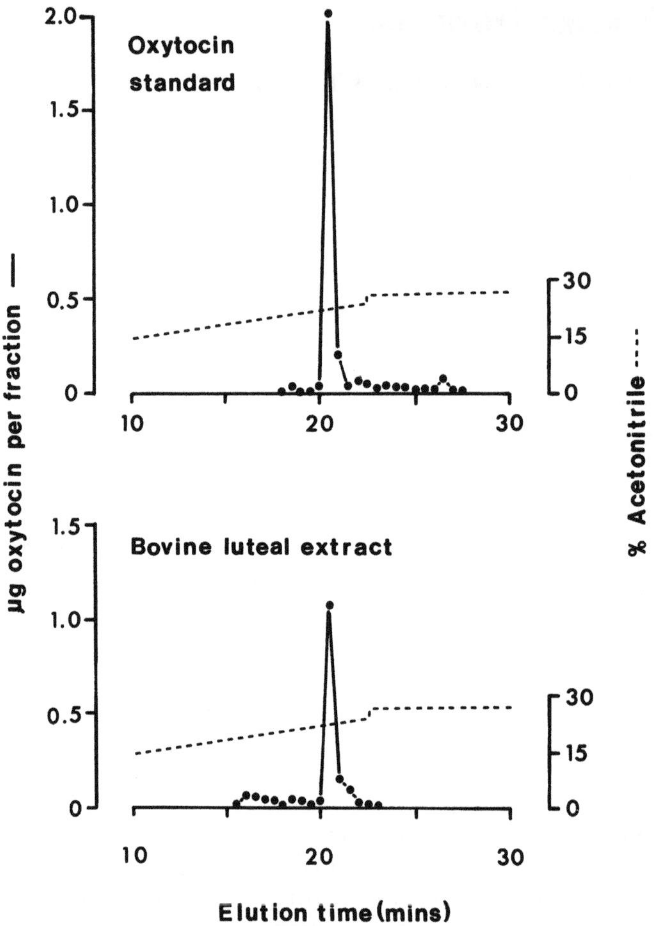

Fig. 1. Profiles obtained by adding bovine luteal extract or oxytocin standard to an HPLC system (Altex model 322) and measuring aliquots of each 0.5-ml fraction by oxytocin radioimmunoassay. (For details of methodology see Wathes and Swann 1982; Wathes et al. 1983 a). The *dashed line* indicates the gradient of acetonitrile

corpus luteum had previously been shown to cause uterine stimulation in the mouse, and the active agent was termed "contractin" (Fields et al. 1980). Fields et al. later showed that the action of contractin could be prevented by preincubation of the extract with a monoclonal antibody to oxytocin (Fields et al. 1983).

As long ago as 1910, Ott and Scott found that an aqueous extract of corpus luteum increased milk flow in the goat by a factor of 4 over a 5-min period, and Schäfer and Mackenzie (1911) and MacKenzie (1911) showed that injection of an extract of ovine corpus luteum could induce milk let-down in the anaesthetized lactating cat. This action has been confirmed by measurements showing an increase in intramammary pressure in rats following the intravenous injection of ovine luteal extract (Wathes and Swann 1982; Wathes et al. 1983 b) or ovine ovarian venous plasma (Sheldrick and Flint 1983 a). In both cases the induced pressure changes were indistinguishable from those caused by treatment with oxytocin.

We have shown that both ovine and bovine luteal extracts dilute in parallel with the oxytocin standard in a radioimmunoassay for oxytocin using two different and specific antisera (Wathes and Swann 1982; Wathes et al. 1983 b). Flint and Sheldrick (1983 a) also measured the luteal compound by radioimmunoassay, and on the basis of the specificity of their antiserum, which recognized both the C-terminal end of the molecule and the ring structure, they concluded that major differences between the luteal compound and oxytocin were unlikely. Parallelism between luteal extract and the oxytocin standard in radioimmunoassay has also been demonstrated by Fields et al. (1983) and Schams et al. (1983 a) in the cow, and Khan-Dawood and Dawood (1983) in the human. In addition, oxytocin has now been identified immunocytochemically in the corpus luteum of the ewe (Watkins 1983) and cow (Guldenaar et al. 1984).

Ovine (Wathes and Swann 1982), bovine (Wathes et al. 1983 a) and human (Wathes et al. 1982) luteal oxytocin extracts all elute at the same position as authentic oxytocin on Sephadex G50 and reverse-phase high-performance liquid chromatography (HPLC) (Fig. 1). These findings have been confirmed by Sheldrick and Flint (1983 a) in the ewe, by Fields et al. (1983) and Schams et al. (1983 a) in the cow and by Schaeffer et al. (1984) in the human. Swann et al. (1984 a) demonstrated that cultures of both ovine and bovine luteal cells will incorporate [^{35}S]-cysteine into a compound which elutes at the same position as oxytocin on HPLC (see following section). If this material is oxidized with performic acid, it produces a new peak coincident with performic acid-oxidized [^{3}H]-oxytocin. Flint and Sheldrick (1985) used fast-atom bombardment mass spectrometry to investigate the molecular mass of the peptide extracted from ovine corpora lutea. The data were consistent with the identification of the ovarian peptide as oxytocin.

Therefore, although the final proof that the ovarian peptide is authentic oxytocin must await the isolation of sufficient material to carry out amino acid sequence analysis, there is now a considerable body of evidence from several different laboratories suggesting that the ovarian and neurohypophyseal hormones are the same.

3 Biosynthesis of Luteal Oxytocin

When immunoreactive oxytocin was first identified in the ovine and then the bovine corpus luteum (see above), it was possible to take the view either that this was due to local synthesis of the hormone within the ovary or that the corpus luteum was acting like a sponge, picking up circulating oxytocin which presumably had its origin in the hypothalamo-neurohypophyseal system. Circumstantial evidence for local synthesis has accumulated from several studies. Flint and Sheldrick (1982) measured an arterial-venous difference in the oxytocin concentration across the ovary but not across the head during the luteal phase of the ewe, in-

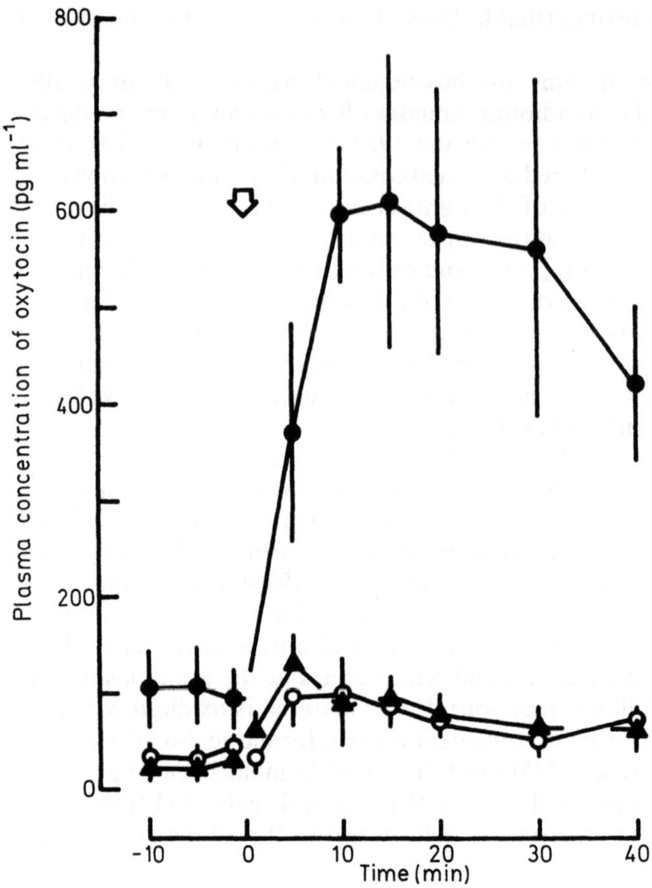

Fig. 2. Oxytocin concentrations measured by radioimmunoassay in plasma from the ovarian vein (●), uterine artery (○) and jugular vein (▲) of four sheep 11–14 days after oestrus. Values are mean ± standard error of mean (SEM). Ovarian venous concentrations from one animal bearing two corpora lutea in the sampled ovary have been halved; thus values are per corpus luteum. Estrumate (125 μg) was administered i.m. at t = 0 (*arrowed*). Anaesthesia was induced with pentobarbitone sodium and maintained with Fluothane in oxygen. The antibody used in the radioimmunoassay showed negligible (<0.01%) cross-reactivity with arginine vasopressin, vasotocin, luteinizing hormone-releasing hormone, ovine luteinizing hormone, progesterone and cloprostenol. (Flint and Sheldrick 1982: reprinted by permission from Nature 297:587, copyright MacMillan Journals Ltd.)

dicating that the ovary was a net secretor of oxytocin. Treatment of luteal-phase ewes with cloprostenol (an analogue of prostaglandin (PG)$F_{2\alpha}$) led to a rapid increase in the oxytocin concentration in the ovarian but not the jugular vein (Fig. 2; Flint and Sheldrick 1982, 1983 b), whereas injection of cloprostenol into ovariectomized ewes did not produce any increase in concentration of circulating oxytocin (Flint and Sheldrick 1983 b). Similarly, in the cow, Walters et al. (1984) found that the mean concentration of oxytocin was higher in the vena cava than in the jugular vein during the luteal phase, again indicating an ovarian rather than a neurohypophyseal source of circulatory oxytocin.

Further indirect evidence for ovarian biosynthesis came from the finding that the bovine corpus luteum also contained a peptide which appeared similar to bovine oxytocin-neurophysin (bNpI). It cross-reacted with bNpI antiserum, showed parallelism in a bNpI radioimmunoassay and eluted at the same position as bNpI on Sephadex G50, G75 and HPLC (Wathes et al. 1983a, b, 1984). Equimolar amounts of oxytocin and neurophysin 1/II have also been measured in ovarian venous plasma in the ewe by Watkins et al. (1983). In addition, recent immunocytochemical studies (Watkins 1983; Guldenaar et al. 1984) have shown staining of the large luteal cells of both sheep and cows with antisera raised to oxytocin and oxytocin-neurophysin, strongly suggesting that some corpora lutea are capable of oxytocin synthesis.

Direct proof of ovarian biosynthesis has been obtained only very recently and has come simultaneously from Richter's laboratory in Hamburg and our own in Bristol. The studies are complementary in that we followed isotope incorporation into dispersed luteal cell cultures to label the proteins, while in Hamburg they started with the luteal mRNA and translated this in a cell-free system. Ivell and Richter (1984) identified the luteal mRNA using oxytocin-specific cDNA probes derived from the cDNA encoding for bovine hypothalamic oxytocin (see chapter by Richter in this volume). Luteal oxytocin mRNA is slightly smaller than its hypothalamic counterpart, judging from their mobility on agarose gels, and this appears to be due to a shorter poly(A) tail on the luteal mRNA. Ivell and Richter also calculated that there is about 250 times more oxytocin mRNA in the most active corpus luteum studied than in a single bovine hypothalamus and discovered, on cell-free translation in a rabbit reticulocyte lysate system, that luteal mRNA gives rise to a protein which is immunoprecipitable by anti-oxytocin-neurophysin and is similar to that derived from hypothalamic mRNA. This protein, which runs on sodium dodecyl sulphate–polyacrylamide gel electrophoresis (SDS-PAGE) gels with an apparent molecular weight of 16500, is therefore the luteal oxytocin precursor. From the cDNA-derived sequence of this precursor, it appears to be identical to its hypothalamic counterpart. Notably, these authors found 1000 times more oxytocin mRNA than vasopressin mRNA in the bovine corpus luteum, and this accords well with the radioimmunoassay data for these hormones (Wathes et al. 1984).

In our laboratory we prepared dispersed cell cultures of bovine and ovine corpora lutea and incubated these with [^{35}S]-cysteine to follow the incorporation of isotope into oxytocin and neurophysin (Swann et al. 1984a). Labelled oxytocin was purified from acid extracts of the cultures by neurophysin-Sepharose affinity chromatography and HPLC, while the neurophysin-like components were immunoprecipitated with antibodies raised to oxytocin (anti-oxy) or rat oxytocin-neurophysin (anti-Np). After 2 h incubation with [^{35}S]-cysteine, a single, 14K, labelled protein was seen on SDS-gels, which could be immunoprecipitated with both anti-oxy and anti-Np. After 6–12 h incubation with isotope, a second 11K–12K, labelled protein was also seen on these gels, but this was recognized only by anti-Np, not by anti-oxy. The 14K protein ran on SDS gels in an identical fashion to rat hypothalamic oxytocin precursor and so represents the luteal oxytocin precursor, while the 11K–12K dalton protein ran identically with neurophysin. The labelling of oxytocin is similar to that of neurophysin, in that no labelled hormone could be detected after incubations of less than 2 h, but was found after 6–12 h.

A major failing in many studies employing isotope incorporation is the lack of controls to discriminate between true incorporation and non-specific binding of labels. This is especially important when [^{35}S]-cysteine is used, as there is the added problem of disulphide interchange. For this reason, the labelled luteal oxytocin was collected after HPLC separation, subjected to performic acid oxidation and rechromatographed by HPLC, where it ran in the position of authentic oxidized oxytocin. Furthermore, it was possible to inhibit incorporation of [^{35}S]-cysteine into oxytocin and neurophysin by treating the cells with the protein synthesis inhibitor, cycloheximide (R.W. Swann, P.J. O'Shaughnessy, S.D. Birkett and B.T. Pickering, unpublished observations).

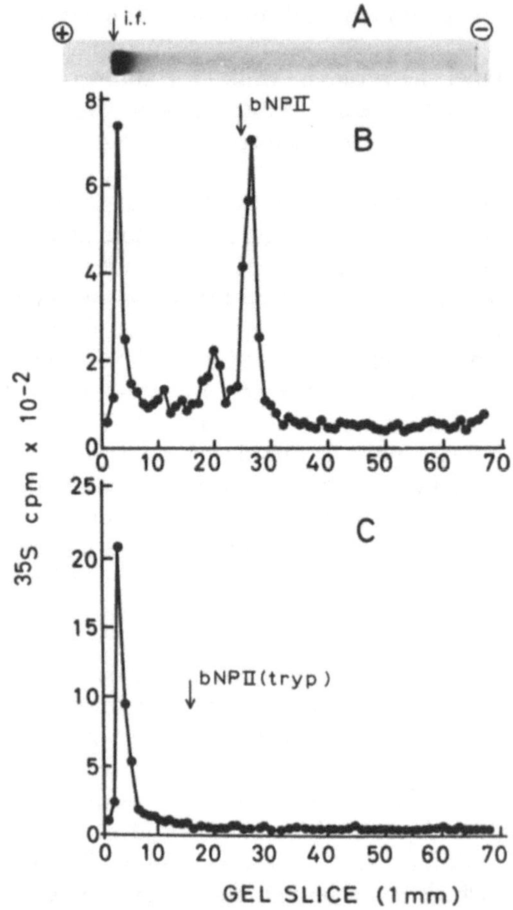

Fig. 3A–C. Basic PAGE of luteal and hypothalamic bNpI. **A** Standard pituitary bNpI, stained with amido-black; **B**, radioactivity profile obtained from a [^{35}S]-labelled anti-neurophysin immunoprecipitate of bovine luteal cells incubated with label for 24 h; **C**, same as **B**, but trypsin-digested for 5 min. *i.f.*, ion front; *bNpII*, bovine pituitary vasopressin-neurophysin; *bNpII (tryp)*, 5-min trypsin-digested bNpII

We were interested to find, along with oxytocin, another labelled molecule which also bound to neurophysin and eluted in the position of vasopressin on HPLC. This radioactive peak was of similar size to the oxytocin peak, but after oxidation moved to a position on HPLC which did not correspond to the oxidized product of either oxytocin or vasopressin. Since the three amino acids at the amino terminus of neurohypophyseal hormones are essential for their binding to neurophysin (see Breslow 1979), it seems likely that this new molecule is related to these hormones. One possibility is that it represents the preamidation form of oxytocin, i.e. oxytocin-Gly-Lys or oxytocin-Gly-Lys-Arg. That it is present in almost equimolar proportions to oxytocin may be simply a reflection of the in vitro environment of the luteal cells, or, on the other hand, it may prove to be of physiological importance.

Peptide-mapping experiments, where we have compared tryptic digests of the reduced and *S*-carboxymethyl derivatives of luteal anti-Np immunoprecipitates with those of standard bovine pituitary oxytocin-neurophysin (bNpI) revealed a close similarity between the two neurophysin sequences (R. W. Swann, S. D. Birkett and B. T. Pickering, unpublished observation). However, it is important to know whether the neurophysin cleaved from luteal oxytocin precursor is the same as bNpI. To test this we first ran luteal anti-Np immunoprecipitates on reverse-phase HPLC, where the major radioactive peak co-eluted with bNpI, and then ran the HPLC peak on a basic PAGE system. On this gel system bNpI runs with or close to the ion front, and has the greatest mobility of all the vertebrate neurophysins that have been analysed in this way (see Cheng and Friesen 1971; Pickering and Jones 1978). Two peaks of radioactivity were obtained, one at the ion front together with bNpI standard, and a second, with an R_F of 0.45, which probably represents luteal oxytocin precursor (Fig. 3). After limited trypsin digestion, all the radioactivity ran with trypsinized bNpI.

It can be concluded from the above that the corpus luteum manufactures oxytocin in a manner similar to that seen in the hypothalamus, i.e. by cleavage of a large precursor protein to yield the hormone and neurophysin. It will be interesting to see whether testicular oxytocin from the Leydig cells (Nicholson et al. 1984; Guldenaar and Pickering 1985) is biosynthesized in the same way and, with the use of dispersed cell cultures, to discover the mechanisms and control of gonadal oxytocin synthesis.

4 Variations in Ovarian Oxytocin Levels

The majority of the work on the characterization of ovarian oxytocin has been performed on luteal tissue collected from non-pregnant cows and sheep. In this section we examine the levels of oxytocin within the ovary at different stages of the oestrous cycle and pregnancy in these species, and look at differences among species.

In both the ewe (Webb et al. 1981; Sheldrick and Flint 1981; Mitchell et al. 1982; Schams et al. 1982) and the cow (Schams 1983; Walters et al. 1984; Walters and Schallenberger 1984), blood oxytocin levels increase during the luteal phase

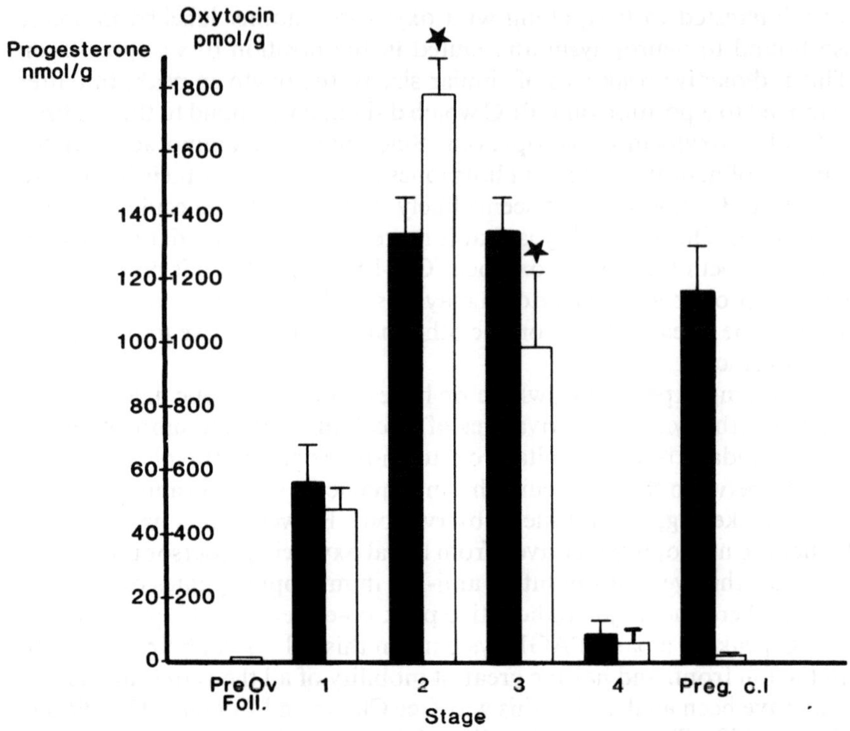

Fig. 4. Progesterone (*solid bars*) and oxytocin (*open bars*) concentrations in corpora lutea collected from cows at different stages of the oestrous cycle and pregnancy. Values are mean ± SEM. *Pre Ov Foll.*, preovulatory follicle ($n=2$); *stage 1*, approximately days 1–4 ($n=6$); *Stage 2*, days 5–10 ($n=7$); *Stage 3*, days 11–17 ($n=10$); *Stage 4*, days 18–20 ($n=6$); *Preg. c.l.*, pregnant days 90–230 ($n=8$). Values marked * are significantly different ($p<0.025$). (Data from Wathes et al. 1984)

and fall during luteolysis to reach basal levels around the time of oestrus. In both species the luteal oxytocin concentration reaches the range of about 1–3 μg/g wet weight of tissue (Wathes and Swann 1982; Sheldrick and Flint 1983 b; Wathes et al. 1984). In the cow we have shown that luteal concentrations are maximal during the first half of the luteal phase and that the decline in oxytocin content apparently occurs before the drop in progesterone content (Fig. 4; Wathes et al. 1984). Similarly, Sheldrick and Flint (1983 b) measured higher levels of luteal oxytocin on day 8 of the ovine cycle than on days 10, 12, and 14 (Fig. 5). These observations are supported by the work of Schams and his co-workers, who found that in both species the decline in circulating oxytocin levels preceded that in circulating progesterone levels (Schams et al. 1983 b; Schams 1983), although Flint and Sheldrick (1983 b) found that plasma levels of both hormones fell simultaneously in the ewe.

Although ovarian oxytocin concentrations are lowest around the time of oestrus we were able to detect some oxytocin in whole bovine follicles (2.3 ± 0.26 ng/g) and Schams et al. (1985) reported values of 42–108 pg/ml in bovine follicular

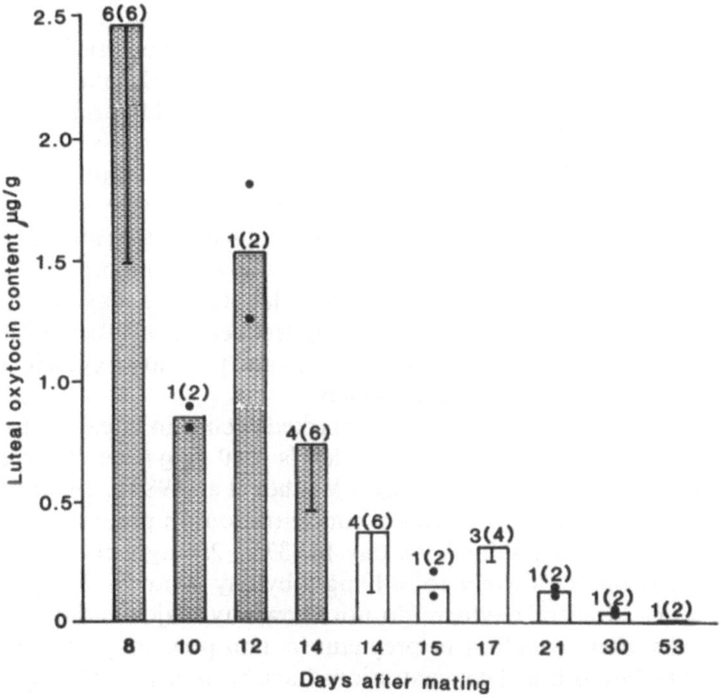

Fig. 5. Oxytocin concentrations in acid extracts of corpora lutea collected from non-pregnant (*hatched bars*) and pregnant (*open bars*) ewes. Animals were mated at oestrus with either intact or vasectomized rams. The number of animals and corpora lutea (*in parenthesis*) per group is shown above. Where this was only two the individual values are also shown, otherwise values are mean ± SEM. (Data from Sheldrick and Flint 1983 b)

fluid. More recently we have measured the oxytocin content in ovine follicles before ovulation. Oxytocin was undetectable during the early follicular phase, appeared at low levels in 3 out of 6 follicles during the LH surge and increased to 26±6 pg/mg in follicle walls and 3.4±0.3 ng/ml in follicular fluid just before ovulation (D.C. Wathes, S.E.F. Guldenaar, R.W. Swann, R.Webb, D.G. Porter, and B.T. Pickering, unpublished observations). It therefore appears that oxytocin production is switched on before ovulation, but increases during the early luteal phase.

Different laboratories are in general agreement about circulating oxytocin levels during the luteal phase but there is a considerable discrepancy in data relating to the follicular phase. Schams et al. (1982) found plasma levels of <3 pg/ml at oestrus. Mitchell et al. (1982), however, found that a pulsatile pattern of oxytocin release persisted during the follicular phase in the ewe with a mean value of just under 40 pg/ml, and Webb et al. (1981) and Flint and Sheldrick (1983 b) could both measure appreciable amounts of oxytocin (18.3±3.4 pg/ml and about 7 pg/ml respectively) at this time.

Our own unpublished data are most similar to those of Flint and Sheldrick (1983 b) in that the mean circulating value was between 5–10 pg/ml around the time of oestrus. Secretion was pulsatile throughout this period, with the height of the pulses decreasing from 25 pg/ml during luteolysis to a minimum of 11 pg/ml during the LH surge. In the cow, Walters and Schallenberger (1984) could detect low-amplitude oxytocin pulses in only two of five animals during the follicular phase, and the mean plasma values for oxytocin in vena cava and jugular vein samples collected at this time were low and similar (1.4 ± 0.2 pg/ml and 1.3 ± 0.4 pg/ml respectively) suggesting that the ovary was not the source. On the basis of their data Mitchell et al. (1982) suggested that an increase in ovarian oxytocin secretion during the luteal phase was superimposed on a background pituitary secretion. The variations among laboratories may be caused by differences in assay methodology and sampling frequency, but they need to be resolved so that the relative contributions of luteal and pituitary oxytocin to circulating levels during the cycle can be determined.

During pregnancy there is a marked decline in luteal oxytocin concentrations in both the sheep and the cow to levels <10 ng/g (ewe, Wathes and Swann 1982, Sheldrick and Flint 1983 b; cow, Wathes et al. 1983 b, 1984). Sheldrick and Flint (1983 b) showed that oxytocin concentrations in pregnant ewes were lower than in non-pregnant animals on day 14 (373 ± 256 ng/g cf. 730 ± 266 ng/g; $n=6$ per group) and fell further to <40 ng/g by day 30 and <5 ng/g by day 53 (Fig. 5). Plasma levels of oxytocin do not show any major variations during the first 2 weeks after ovulation in pregnant vs non-pregnant ewes (Sheldrick and Flint 1981), but as luteal concentrations have been measured only from day 14, when there was already an apparent difference, it remains to be determined how early in pregnancy the difference is manifest. However, Schams and Lahlou-Kassi (1984) found that plasma levels decreased about two days earlier in pregnancy than during the oestrous cycle.

A similar pattern of ovarian oxytocin production probably occurs in the goat, in which circulating oxytocin levels fall from >40 pg/ml to <10 pg/ml between days 12 and 16 in both pregnant and non-pregnant animals (Homeida and Cooke 1983), and become undetectable following ovariectomy (Cooke et al. 1984). The same authors have extracted oxytocin-like material from caprine corpora lutea which has both immuno- and bioactivity comparable to that of oxytocin and is present at a concentration of about 1.3 µg/g wet weight of tissue (A. M. Homeida and R. G. Cooke, personal communication).

Apart from ruminants, the only other group which has been studied in any detail is the primates, in particular the human female. Ovarian tissue collected from women at various stages of the menstrual cycle was pooled, and immunoreactivity eluting at the same position as the oxytocin standard was detected following purification on Sephadex G50 and HPLC (Wathes et al. 1982). The overall concentration was estimated at 30 ng/g. Since then we have measured oxytocin concentrations in eight individual samples of ovarian tissue collected at various stages of the cycle, and in 12 samples of fluid from preovulatory or cystic follicles (D. C. Wathes, P. G. Wardle and M. G. R. Hull, unpublished observations). Oxytocin levels were at or below the limit of detection of the assay in all but one sample of follicular fluid, which contained 75 pg/ml. The ovarian concentrations (in whole follicles, wedges of ovary or corpora lutea) were lower than those previously recorded and ranged from <0.03 to 5.4 ng/g. They showed no obvious correlation with the stage of the cycle, which was estimated from menstrual history and steroid levels in plasma taken at the time of tissue collection. However, the majority of the women donors were undergoing hysterectomy following a history of irregular cycles.

Two similar studies have been performed by other groups. Khan-Dawood and Dawood (1983) measured oxytocin by radioimmunoassay in extracts of ovarian tissue from 26 women. They reported levels of 11–53 ng/g in corpora lutea ($n=7$), 1.4–60 ng/g in preovulatory follicles ($n=3$) and 0–21 ng/g in ovarian stroma ($n=5$). A corpus luteum of pregnancy in this study contained 106 ng/g, whereas in our study a similar sample contained 0.2 ng/g. Schaeffer et al. (1984) found oxytocin concentrations of 3.1 ng/mg protein in corpora lutea ($n=3$), 0.48 ng/mg protein in whole ovaries ($n=9$) and 299 pg/ml in follicular fluid ($n=3$) analysis. The evidence available at present therefore suggests that human ovaries do contain immunoreactive oxytocin. However, concentrations are about three orders of magnitude lower than those found in ruminant corpora lutea and do not show any obvious correlation with the stage of the cycle or with pregnancy, although this may in part reflect the quality of the material available.

Khan-Dawood et al. (1984) have also detected immunoreactive oxytocin in corpora lutea obtained from non-pregnant cynomolgus monkeys. The concentration ranged from 3.4 to 602.5 ng/g wet weight of tissue. Measurements of oxytocin in blood showed that the concentration in ovarian venous plasma on days 15–24 of the cycle was significantly higher than that in peripheral venous blood or ovarian venous blood collected on days 26–28 (161 ± 29 pg/ml vs 16.8 ± 4 pg/ml and 23.2 ± 3 pg/ml respectively). This suggests that oxytocin is secreted by the corpus luteum during the main part of the luteal phase. In our laboratory, however, we have been unable to detect oxytocin in marmoset ovaries. The concentration was <60 pg/ovary in all eight ovaries taken from four animals at various stages of the cycle (A. F. Dixson and D. C. Wathes, unpublished observations).

There is also a marked difference between primates and ruminants in the pattern of peripheral oxytocin levels during the cycle. Whereas in ruminants circulating levels are maximal in the early luteal phase, a number of studies on women and rhesus monkeys have indicated a rise in circulating oxytocin or neurophysin lasting for about 4 days around the time of ovulation. There may be a second minor increase during the luteal phase, but levels during the rest of the menstrual cycle tend to be very variable (human, Legros et al. 1975; Mitchell et al. 1980; Amico et al. 1981; monkey, Robinson et al. 1976; Falconer et al. 1980). So although the ovarian oxytocin concentrations in women are higher than those in plasma, which are in the range of 1–10 pg/ml (Amico et al. 1981), there is no evidence at present that ovarian secretion is responsible for alterations in the peripheral concentration of the hormone.

The only other species we have examined is the pig. Oxytocin was present in an extract of non-pregnant sow corpus luteum, but the concentration was extremely low (0.25 ng/g wet weight of tissue). This figure is in close agreement with the work of Pitzel et al. (1984) who measured the ovarian oxytocin concentration at three stages during the oestrous cycle. On days 12 and 19 the concentrations ranged from 0.3 to 0.8 ng/g ($n=4$) whereas the values were apparently higher on day 5 (6.4–10.8 ng/g, $n=3$). Therefore it appears that the pig is another species in which the concentration is maximal during the early luteal phase, although the total amount present is much less than that found in ruminants. Further work is therefore required on a wider range of species to determine whether high concentrations of luteal oxytocin are a phenomenon confined to ruminants or whether they also occur in other mammals.

5 Control of Synthesis and Secretion

The majority of the data available again refer to the ewe and the cow, with limited information on the goat. In these species circulating oxytocin concentrations increase after oestrus, with a maximum in the luteal phase and a decline during luteolysis (see Sect. 4). In the ewe oxytocin is first detectable by radioimmunoassay of follicle wall extracts during or shortly after the LH surge. Immunocytochemical techniques have shown that this oxytocin is located in the granulosa cells (D. C. Wathes, S. E. F. Guldenaar and B. T. Pickering, unpublished observations) and similar findings have recently been reported for the cow (Kruip et al. 1985). In the ewe the concentration of oxytocin remains low (<100 ng/g) until at least 12 h after ovulation, but increases to luteal levels (>1 μg/g) within three days. However our immunocytochemical evidence suggests that the amount of precursor present may increase immediately after ovulation. This implies that the initial stimulus to begin production occurs during or shortly before the LH surge, but that initially only some of the precursor is processed to oxytocin. Possible endocrine factors which may regulate synthesis and processing of the hormone by the granulosa cells are currently under investigation.

In the mature corpus luteum of the non pregnant ewe (Watkins 1983) and cow (Guldenaar et al. 1984) specific staining for both oxytocin and neurophysin has been demonstrated in the large luteal cells which are known to contain secretory granules (Gemmel et al. 1977; Parry et al. 1980). The exocytos of these granules accelerates during the mid-luteal phase, prompting the suggestion that they contain progesterone (Gemmel et al. 1977; Gemmel and Stacy 1979; Quirk et al. 1979) but it now seems more likely that the granules represent the luteal oxytocin store. This view is supported by two studies. Rice and Thorburn (1985) showed that the oxytocin in ovine corpora lutea was associated with a particulate fraction sedimenting at a density of 1.054–1.061 g/ml which contained electron dense granules with a diameter of 200–250 nm, and Theodosis et al. (1985) showed specific staining of similar granules with both oxytocin and neurophysin antisera using an immunogold labelling technique.

After about two weeks the luteal oxytocin content declines again in both pregnant and non-pregnant ewes (see above). This initially suggested a pregnancy specific inhibition of oxytocin synthesis, but Sheldrick and Flint (1983c) found that

Fig. 6 A–D. Alternate sections of the corpus luteum of a non-pregnant cow stained with **A** anti-oxytocin, **B** anti-vasopressin, **C** anti-oxytocin-neurophysin and **D** anti-vasopressin-neurophysin. *BV*, blood vessel. (Guldenaar et al. 1984)

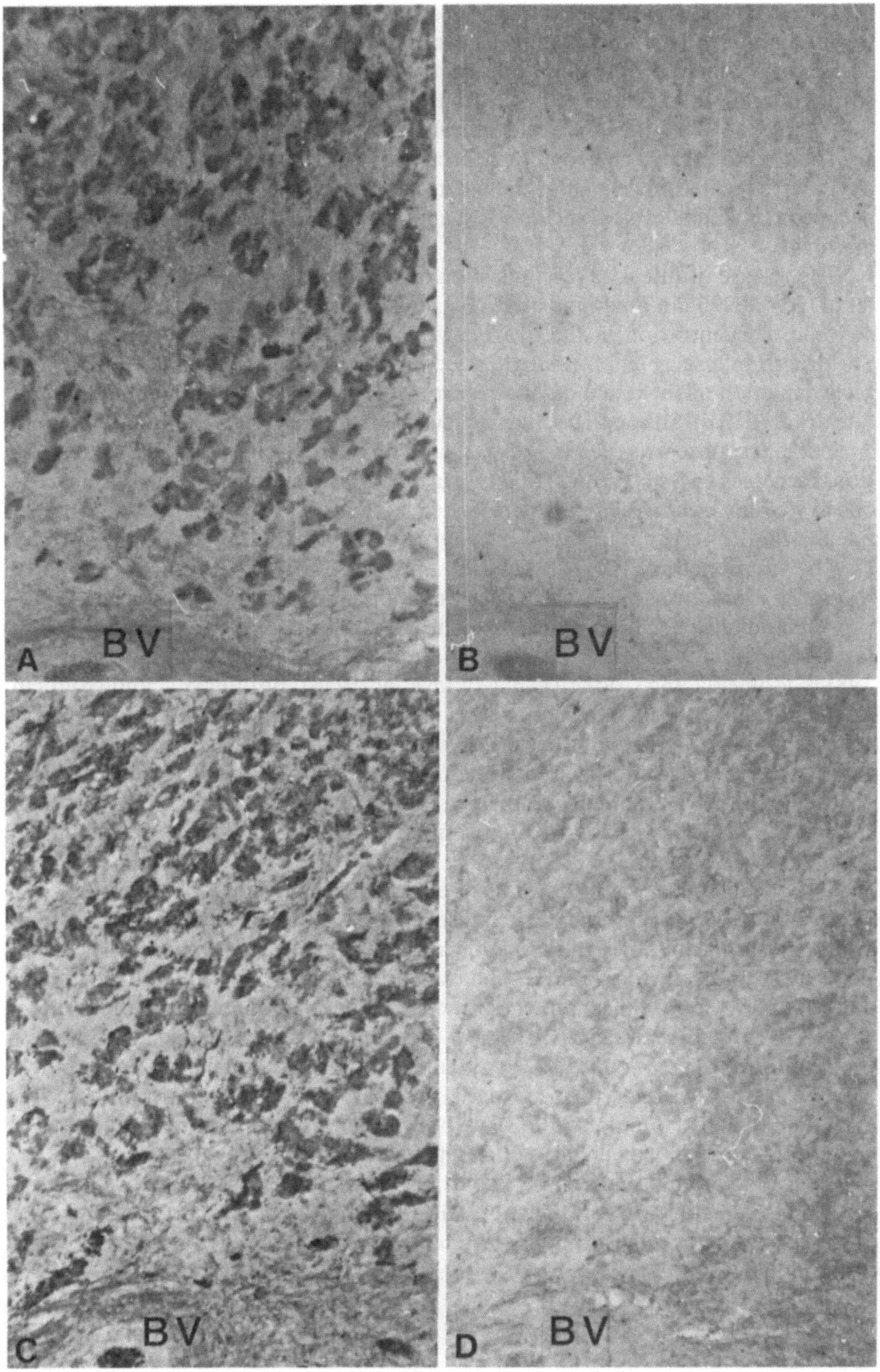
A
BV
B
BV
C
BV
D
BV

the luteal oxytocin concentration also dropped at the same time in ewes in which the lifespan of the corpus luteum was prolonged by hysterectomy. Therefore the decline in oxytocin production appears to be related to an ageing process in the corpus luteum. It is of interest that Alila and Hansel (1984) recently demonstrated that in the cow the large luteal cells initially develop from the granulosa layer but are soon replaced by the growth of theca-derived cells. As we have shown that oxytocin is located in cells of granulosa origin in the ewe, it seems likely that the decline in oxytocin levels coincides with a loss of this cell line as the corpus luteum matures.

The release of luteal oxytocin is stimulated by the administration of cloprostenol (ICI-80996, an analogue of $PGF_{2\alpha}$) which causes a marked increase in the oxytocin concentration in the ovarian vein within 5 min in the ewe (Flint and Sheldrick 1982; see Fig. 2). A single injection of 125 µg i.m. caused almost complete depletion of the luteal oxytocin store within 45 min of treatment (Flint and Sheldrick 1983 b), although circulating levels did not become basal for about 24 h. A similar response was seen in the cow, in which circulating oxytocin concentrations increased to a maximum 5–10 min after cloprostenol treatment and then declined steadily over a 3- to 5-h period, reaching a minimum 24–36 h after treatment (Schallenberger et al. 1984). The release of ovarian oxytocin in the ewe can also be stimulated by PGE_2 (Sheldrick and Flint 1983 a). Several groups have measured circulating levels of $PGF_{2\alpha}$ or its metabolite 13, 14-dihydro-15-keto-prostaglandin F (PGFM), and oxytocin or its associated neurophysin during luteal regression in the sheep and goat (Fairclough et al. 1980; Flint and Sheldrick 1983 b; Homeida and Cooke 1983; Cooke et al. 1984). Both $PGF_{2\alpha}$ and oxytocin are released in a pulsatile fashion, with a tendency for the pulses to coincide. Oxytocin is known to stimulate the release of $PGF_{2\alpha}$ from the uterine endometrium in the ewe (Roberts and McCracken 1976). Hence it appears that during luteolysis there is a positive feedback loop in operation between ovarian oxytocin and uterine $PGF_{2\alpha}$ which leads to the eventual depletion of the ovarian oxytocin store.

In a detailed study of hormone secretion during the luteal phase of the cow, Walters et al. (1984) compared the pulsatile release pattern of progesterone and oxytocin during the early (day 4) and mid- (days 9–14) luteal phases. The basal concentration of oxytocin in the vena cava and the pulse amplitude for oxytocin were similar at both times, but the pulse frequency was higher in the mid-luteal phase. In both phases nearly 100% of oxytocin pulses coincided with progesterone pulses, whereas only 29% (early) and 86% (mid) of all progesterone pulses were associated with oxytocin release. It is possible that both hormones are secreted in response to pulses of $PGF_{2\alpha}$, but that additional factors stimulate progesterone secretion, particularly during the early luteal phase.

In conclusion, all that can be said at present is that exogenous $PGF_{2\alpha}$ stimulates the release of ovarian oxytocin. All other factors relating to the control of ovarian oxytocin synthesis and secretion await elucidation.

6 Actions of Ovarian Oxytocin

Two main actions have so far been attributed to ovarian oxytocin. These are an involvement in luteolysis which in ruminants is mediated via the uterus and a direct effect on steroidogenesis. Oxytocin is also known to stimulate motility of the reproductive tract, and it may influence gonadotrophin secretion. However, for reasons which have been discussed in a recent review (Wathes 1984), it is less likely that oxytocin from an ovarian source is of importance in these last two functions, and they will not be considered further here.

The luteolytic action of oxytocin was first described by Armstrong and Hansel (1959), who found that daily injections of 50–150 units oxytocin in heifers during the 1st week of the oestrous cycle led to a significant decrease in oestrous cycle length. Hansel and Wagner (1960) then reported that oxytocin was only luteolytic between days 3 and 6 of the cycle, treatments at other times proving ineffective. These data were later confirmed by other workers (e.g. Labhsetwar et al. 1964; Donaldson and Takken 1968; Harms et al. 1969). In the goat, oxytocin treatment on days 3–6 of the cycle also led to a premature return to oestrus (Cooke and Knifton 1981; Cooke and Homeida 1982). Exogenous oxytocin, however, does not appear to cause luteal regression in ewes, although it may be followed by a slight depression in progesterone secretion (Milne 1963; Hatjiminaoglou et al. 1979; Duby et al. 1983), and both active and passive immunization against oxytocin led to an increase in oestrous cycle length in this species (Sheldrick et al. 1980; Schams et al. 1983b). In the rhesus monkey intramuscular injection of a large dose of oxytocin during the luteal phase did not alter luteal function (Wilks 1983), whereas a much lower dose infused directly into the corpus luteum caused a significant decrease in progesterone secretion and reduced the length of the luteal phase (Auletta et al. 1984). In a variety of other species exogenous oxytocin has so far proved ineffective in reducing oestrous cycle length (horse, Neely et al. 1979; pig, Duncan et al. 1961; rat and rabbit, Brinkley and Nalbandov 1963; guinea pig, Donovan 1961).

In the three ruminant species which have been studied, oxytocin treatment is known to stimulate the release of $PGF_{2\alpha}$ from the uterine endometrium (sheep, Roberts and McCracken 1976; goats, Cooke and Homeida 1982; heifers, Newcomb et al. 1977, Milvae and Hansel 1980). It appears that the luteolytic action of oxytocin is normally mediated by this response, since it can be prevented by hysterectomy in the cow (Armstrong and Hansel 1959; Anderson et al. 1965; Ginther et al. 1967) and by simultaneous treatment with meclofenamic acid (a prostaglandin synthetase inhibitor) in the goat (Cooke and Knifton 1981; Cooke and Homeida 1983).

The concentration of endometrial oxytocin receptor changes during the oestrous cycle of the ewe, reaching a maximum at oestrus and declining to an almost undetectable level during the mid-luteal phase before increasing again on day 15 of the cycle (Roberts et al. 1976). The appearance of the receptors is apparently controlled by circulating steroid levels (McCracken et al. 1978). In the ewe and the cow, $PGF_{2\alpha}$ is not luteolytic during the first few days of the cycle (Rowson et al. 1972; Haresign 1978). The period between days 3 and 6 when oxytocin does induce luteolysis in cows and goats is therefore presumed to reflect a time when sufficient receptors of the appropriate type are present for oxytocin to act on the endometrium and for $PGF_{2\alpha}$ to act on the corpus luteum.

Such a situation may not occur in the ewe. Nevertheless, it has been postulated by McCracken and colleagues that endogenous oxytocin does play an important role in natural luteolysis in this species (see McCracken 1980; McCracken et al. 1981). These workers have demonstrated an increase in the endometrial oxytocin receptor on day 15 of the cycle (Roberts et al. 1976), and have suggested that this is the triggering event for luteolysis. However, they did not measure plasma steroid concentrations, so it cannot be certain whether the increase in receptor concentration precedes or follows the fall in progesterone. Measurements of oxytocin and $PGF_{2\alpha}$ in the peripheral circulation during luteolysis support the idea that oxytocin is involved in this process, as both hormones exhibit pulsatile release during this period, with a tendency for the pulses to coincide (see Sect. 5). It was originally assumed that the oxytocin originated from the neurohypophysis, but luteal oxytocin concentrations decline during luteolysis, suggesting that the ovary is a more likely source.

To summarize: ovarian oxytocin is secreted during luteolysis in ruminants, and this stimulates the release of endometrial $PGF_{2\alpha}$, which in turn decreases luteal progesterone production and also causes the release of further oxytocin. It is not yet clear whether the appearance of endometrial oxytocin receptors triggers this chain of events or follows a fall in progesterone secretion induced by some other factor. Exogenous oxytocin does not appear to be luteolytic in a variety of other species.

Evidence for a direct action of oxytocin on steroidogenesis in the female is less conclusive. Tan et al. (1982a, b) carried out two studies using short-term incubations of dispersed luteal cells from the cow and human. In both cases there was a tendency for lower doses of oxytocin (4–40 mU/ml) to stimulate progesterone production whereas higher doses (> 400 mU/ml) inhibited both basal and hCG-stimulated progesterone release. However, Richardson, and Masson (1985) were unable to demonstrate such an effect using human luteal cells, and Mukhopadhyay et al. (1984) tested a range of oxytocin concentrations from 10^{-8} to 4×10^{-6} M on luteal cells from pseudopregnant rats and found no alteration of either basal or hCG-stimulated progesterone production.

The early study by Mares and Casida (1963) did not reveal an influence of oxytocin on progesterone production by slices of bovine luteal tissue. We have used isolated bovine luteal cells, collected from either cyclic or pregnant cows and maintained in vitro for 3- or 12-h periods. In each experiment the effect of a range of oxytocin levels from 10^{-11} to 10^{-6} M was tested on hCG-stimulated cells. In individual experiments there was some variation in progesterone production with different doses of oxytocin, but there was no consistent trend, and we were therefore unable to confirm the data of Tan et al. (1982a) (J. Barrett, P. J. O'Shaughnessy and D. C. Wathes, unpublished observations).

Several situations have been demonstrated where pronounced alterations in oxytocin levels are not accompanied by any obvious variation in progesterone output in vivo. Thus, luteal oxytocin levels fall sharply at about day 14 in ewes in which progesterone secretion is maintained by either pregnancy or hysterectomy (Sheldrick and Flint 1983 b), and progesterone also remains at luteal levels in the plasma of ewes in which the cycle is prolonged by active immunization against oxytocin (Schams et al. 1983 b).

Support for the idea that oxytocin may influence steroidogenesis in the male has come from the work of Adashi and Hsueh (1981 a, b). They demonstrated from cultures of rat testicular cells that oxytocin, vasopressin and vasotocin could all cause a dose-dependent inhibition of hCG-stimulated testosterone secretion. Further studies using vasotocin suggested that the inhibition was due to the selective suppression of the enzymes 17α-hydroxylase and 17–20 desmolase, thus reducing conversion of progesterone to androstenedione (Adashi and Hsueh 1982). If oxytocin has a similar site of action in the female it is unlikely to have a major influence on progesterone secretion, but could affect ovarian production of androgens and oestrogens. Further studies are required, therefore, before a direct action of oxytocin on progesterone secretion at the level of the corpus luteum can be confirmed.

7 Oxytocin in the Male

Acid extracts of testes from a number of mammalian species contain immunoreactive oxytocin (human, rat, mouse, dog, cat, rabbit, sheep, bull: Nicholson et al. 1984, Wathes 1984, Pickering et al. 1986, Schams et al. 1985). In our laboratory the mean concentration in these species ranged from 0.1 ng/g (ram) to 1.0 ng/g (rat), and so the male ruminants do not share with the females the characteristic of having large amounts of oxytocin in their gonads. In both the rat and human we have shown that the immunoreactive oxytocin dilutes in parallel with the standard curve, and elutes at the same position as oxytocin on HPLC (Nicholson et al. 1984). We have not yet been able to demonstrate incorporation of radiotracer into oxytocin by testicular tissue, but in both the rat and the human neurophysin is also present (Nicholson et al. 1984) and preliminary experiments using cultures of dispersed human testicular cells have been shown to secrete oxytocin into the medium over an 11 day period. Immunocytochemical studies have shown that 80% of interstitial cells stain for oxytocin, whereas no staining was seen in the tubules (Guldenaar and Pickering 1985). Thus it seems reasonable to suggest that oxytocin is located in the Leydig cells.

In the testis the seminiferous tubules have a myoid layer which is responsible for their contractile activity. When rat seminiferous tubules were perifused and observed with videomicrography they showed two sorts of movement, localised dimpling and pronounced contractions which resulted in the propulsion of the contents of the lumen along the tubule (Worley et al. 1985). Both of these types of movements were stimulated by small doses (ca. 1 ng/ml) of oxytocin. Several studies we have carried out suggest that testicular oxytocin is normally responsible for tubule motility. In the rat the oxytocin content varies with age, and is first detectable in the newborn rat on day 7, which corresponds with a marked increase in the contractility of the seminiferous tubules in vitro (Worley et al. 1985). Treatment of male rats with ethane-1,2, dimethanesulphonate leads to a temporary disappearance of the Leydig cells. Three days after such treatment immunoreactive oxytocin could no longer be detected in the testes and motility was dramatically reduced. Within 28 days both the oxytocin content and motility had returned to normal (Pickering et al. 1986). In the hypogonadal mouse (a mutant strain) the Leydig cells remain atrophic because of a lack of circulating LH, and testosterone production is also negligible. In the testes of such mice oxytocin is undetectable and the tubules do not contract. Treatment with LH for 15 days produced some oxytocin and some tubular movement, whereas testosterone implants resulted in normal oxytocin concentrations and tubule motility (Pickering et al. 1986). Therefore it appears that tubular activity is related to oxytocin content, and that this may in turn be related to androgen production.

Nicholson et al. (1985) have also identified an oxytocin-like peptide in human prostatic tissue and seminal fluid at concentrations of 1.8 ng/g and 1.5 ng/ml respectively. However Schams et al. (1985) found that oxytocin was undetectable (<5 pg/g) in bull seminal vesicles and prostate glands.

8 Other Sources of Oxytocin

Excluding the brain and certain tumours, high concentrations of oxytocin have been reported only in the ovary. However, several other glandular tissues have been investigated and have been shown to contain oxytocin in concentrations considerably greater than those in blood. There have been two reports of oxytocin immunoreactivity in human placental tissue. Fields et al. (1983) examined term placentae and found that they contained 3 ng oxytocin/mg (but did not state whether this was wet or dry weight). This material showed parallelism with the oxytocin standard curve and re-initiated relaxin-inhibited contractions of the mouse uterus in vitro. The bioactivity eluted at the appropriate position for oxytocin on HPLC. Makino et al. (1983) also obtained a human placental extract with immuno- and bioactivity similar to that of oxytocin. The concentration increased from 1.6 ± 0.3 ng/g tissue in the 1st trimester to 20.8 ng/g in the 2nd trimester, with a similar concentration at term. They estimated that the total amount of oxytocin present in the 3rd trimester was 12.5 ± 1.2 µg per placenta, an amount similar to that in the posterior pituitary (see Lederis and Jayasena 1970).

Several groups have now identified oxytocin in the adrenal gland. Nicholson et al. (1984) reported the presence of both immunoreactive oxytocin and neurophysin in a pool of adrenal tissue collected from human fetuses at various stages of gestation and Ang and Jenkins (1984) found oxytocin and neurophysin in the adrenal medulla of both the adult human and the rat. Both studies showed that the peptides eluted at the appropriate position on HPLC. Schams et al. (1985) found that oxytocin was present in cow adrenals, with significantly higher concentrations in the medulla than the cortex (713 versus 72 pg/g).

Ott and Scott (1910, 1911) found that an aqueous extract of thymus gland (species not stated) increased milk let-down in the lactating goat. We obtained an acidic extract of 16-week veal calf thymus. A small amount of immunoreactive oxytocin (0.2 ng/g) eluted at the appropriate position on a column of Sephadex G75. No oxytocin was found in a sample of lamb thymus, but at the animals from which the tissue was obtained were nearly a year old the activity of the gland may have declined.

Thus, oxytocin-like activity has been identified at sites in the body other than the neurohypophysis and gonads. The concentrations are generally quite low, and although they could represent tissue uptake, the additional finding that neurophysin is present in the testis and in the adrenal suggests that local biosynthesis of oxytocin may occur. However, the function of oxytocin at these other sites remains undetermined.

9 Conclusion

Compelling evidence has now been accumulated that oxytocin is produced in the ovaries of the cow, the ewe and the goat. Synthesis begins in the granulosa layer of the preovulatory follicle. When luteinization occurs oxytocin production increases and continues in the large cells of the corpus luteum, reaching maximal levels during the first half of the oestrous cycle. Oxytocin levels in the corpus luteum decline after about two weeks even if the lifespan is prolonged by pregnancy or hysterectomy. In the non-pregnant animals pulses of oxytocin which are secreted during luteolysis are probably important in stimulating the release of PG $F_{2\alpha}$ from the uterine endometrium. It has also been suggested that oxytocin may have a direct influence on luteal progesterone synthesis, but not all reports are in agreement. As production is maximal during the early part of the cycle it seems likely that ovarian oxytocin plays an additional role at this time which has not yet been elucidated.

Oxytocin is also present in the ovaries of some primates and in pigs, but the concentrations are much lower than those found in ruminants. In the human there is no obvious cyclical variation of the amount of peptide, and its function is unknown. A wider range of species needs to be studied in order to determine whether large amounts of ovarian oxytocin are to be found only in ruminants, and to determine the significance of the variation between species.

Oxytocin immunoreactivity has also been demonstrated in a variety of other glands which now include the testes, adrenals, prostate, placenta and thymus from several different species. At all of these sites the concentration is much lower than that found in the ruminant corpus luteum, but still exceeds blood levels. Although direct evidence for biosynthesis has not yet been obtained for any of these locations, several also contain immunoreactive neurophysin which suggests that production may occur at sites other than the ovary. In the testis evidence is accumulating that oxytocin is involved in sperm transport by stimulating contractions of the seminiferous tubules, but its role elsewhere is unknown. However it is becoming clear that the production of oxytocin is not confined to the hypothalamus and its presence elsewhere in the body suggests that it may have a wider range of physiological actions than has previously been recognized.

Acknowledgements. Work in the authors' laboratories was supported in part by the MRC, AFRC and the Wellcome Trust.

References

Adashi EY, Hsueh AJW (1981 a) Direct inhibition of testicular androgen biosynthesis revealing antigonadal activity of neurohypophysial hormones. Nature 293:650–652

Adashi EY, Hsueh AJW (1981 b) Direct inhibition of testicular androgen biosynthesis by arginine-vasopressin: mediation through pressor-selective testicular recognition sites. Endocrinology 109:1793–1795

Adashi EY, Hsueh AJW (1982) Direct inhibition of rat testicular androgen biosynthesis by arginine vasotocin. Studies on mechanisms of action. J Biol Chem 257:1301–1308

Alila HW, Hansel W (1984) Origin of different cell types in the bovine corpus luteum as characterized by specific monoclonal antibodies. Biol Reprod 31:1015–1025

Amico JA, Seif SM, Robinson AG (1981) Elevation of oxytocin and oxytocin-associated neurophysin in the plasma of normal women during mid-cycle. J Clin Endocrinol Metab 53:1229–1232

Anderson LL, Bowerman AM, Melampy RM (1965) Oxytocin on ovarian function in cycling and hysterectomized heifers. J Anim Sci 24:964–968

Ang VTY, Jenkins JS (1984) Neurohypophysial hormones in the adrenal medulla. J Clin Endocrinol Metab 58:688–691

Armstrong DT, Hansel WJ (1959) Alteration of the bovine estrous cycle with oxytocin. J Dairy Sci 42:533–542

Auletta FJ, Paradis DK, Wesley M, Duby RT (1984) Oxytocin is luteolytic in the rhesus monkey (*Macaca mulatta*). J Reprod Fertil 72:401–406

Breslow E (1979) Chemistry and biology of the neurophysins. Ann Rev Biochem 48:251–271

Brinkley HJ, Nalbandov AV (1963) Effect of oxytocin on ovulation in rabbits and rats. Endocrinology 73:515–517

Cheng KW, Friesen HG (1971) Isolation and characterization of a third component of porcine neurophysin. J Biol Chem 246:7656–7665

Cooke RG, Homeida AM (1982) Plasma concentrations of 13,14-dihydro-15-keto prostaglandin $F_{2\alpha}$ and progesterone during oxytocin-induced oestrus in the goat. Theriogenology 18:453–460

Cooke RG, Homeida AM (1983) Prevention of the luteolytic action of oxytocin in the goat by inhibition of prostaglandin synthesis. Theriogenology 20:363–365

Cooke RG, Knifton A (1981) Oxytocin-induced oestrus in the goat. Theriogenology 16:95–97

Cooke RG, Homeida AM, Watkins WB (1984) Simultaneous release of neurophysin and ovarian oxytocin during luteolysis in the goat. J Physiol 354:97 p

Donaldson LE, Takken A (1968) The effect of exogenous oxytocin on corpus luteum function in the cow. J Reprod Fertil 17:373–383

Donovan BT (1961) The role of the uterus in the regulation of the oestrous cycle. J Reprod Fertil 2:508–510

Duby RT, Tritschler JP, Pekala RF, Gnatek GJ, Dubois D, Perry R (1983) Effect of oxytocin and sodium ibuprofenate on corpus luteum function in the ewe. J Anim Sci [Suppl 1] 57:330 (abstract)

Duncan GW, Bowerman AM, Anderson LL, Hearn WR, Melampy RM (1961) Factors influencing in vitro synthesis of progesterone. Endocrinology 68:199–207

Fairclough RJ, Moore LG, McGowan LT, Peterson AJ, Smith JF, Tervit HR, Watkins WB (1980) Temporal relationship between plasma concentrations of 13,14-dihydro-15-keto-prostaglandin F and neurophysin I/II around luteolysis in sheep. Prostaglandins 20:199–208

Fairclough RJ, Moore LG, Peterson AJ, Watkins WB (1984) Effect of oxytocin on plasma concentrations of 13,14-dihydro-15-keto prostaglandin F and the oxytocin-associated neurophysin during the estrous cycle and early pregnancy in the ewe. Biol Reprod 31:36–43

Falconer J, Mitchell MD, Moutford LA, Robinson JS (1980) Plasma oxytocin concentrations during the menstrual cycle in the rhesus monkey, *Macaca mulatta*. J Reprod Fertil 59:69–72

Fields MJ, Fields PA, Castro-Hernandez A, Larkin LH (1980) Evidence for relaxin in corpora lutea of late pregnant cows. Endocrinology 107:869–876

Fields PA, Eldridge RK, Fuchs AR, Roberts RF, Fields MJ (1983) Human placental and bovine corpora luteal oxytocin. Endocrinology 112:1544–1546

Flint APF, Sheldrick EL (1982) Ovarian secretion of oxytocin is stimulated by prostaglandin. Nature 297:587–588

Flint APF, Sheldrick EL (1983a) Secretion of oxytocin by the corpus luteum in sheep. Prog Brain Res 60:521–530

Flint APF, Sheldrick EL (1983b) Evidence for a systemic role for ovarian oxytocin in luteal regression in sheep. J Reprod Fertil 67:215–225

Flint APF, Sheldrick EL (1985) Ovarian oxytocin. In: Amico JA, Robinson AG (eds) Oxytocin: Clinical and laboratory studies. Elsevier, BV, pp 335–350

Gemmel RT, Stacy BD (1979) Effect of cycloheximide on the ovine corpus luteum: the role of granules in the secretion of progesterone. J Reprod Fertil 57:87–89

Gemmel RT, Stacy BD, Nancarrow CD (1977) Secretion of granules by the luteal cells of the sheep and the goat during the estrous cycle and pregnancy. Anat Rec 189:161–168

Ginther OJ, Woody CO, Mahajan S, Janakiraman K, Casida LE (1967) Effect of oxytocin administration of the oestrous cycle of unilaterally hysterectomized heifers. J Reprod Fertil 14:225–229

Guldenaar SEF, Pickering BT (1985) Immunocytochemical evidence for the presence of oxytocin in rat testis. Cell Tissue Res 240:485–487

Guldenaar SEF, Wathes DC, Pickering BT (1985) Immunocytochemical evidence for the presence of oxytocin and neurophysin in the large cells of the bovine corpus luteum. Cell Tissue Res 237:349–352

Hansel W, Wagner WC (1960) Luteal inhibition in the bovine as a result of oxytocin injections, uterine dilatation, and intrauterine infusions of seminal and preputial fluids. J Dairy Sci 43:796–805

Haresign W (1978) Ovulation control in the sheep. In: Crighton DB, Foxcroft GR, Haynes NB, Lamming GE (eds) Control of ovulation. Butterworths, London, pp 435–451

Harms PG, Niswender GD, Malven PV (1969) Progesterone and luteinizing hormone secretion during luteal inhibition by exogenous oxytocin. Biol Reprod 1:228–233

Hatjiminaoglou I, Alifakiotis T, Zervas N (1979) The effect of exogenous oxytocin on estrous cycle length and corpus luteum cysts in ewes. Ann Biol Anim Biochem Biophys 19:355–365

Homeida AM, Cooke RG (1983) Suppression of pulsatile release of oxytocin during early pregnancy in the goat. Prostaglandins 26:103–109

Ivell R, Richter D (1984) The gene for the hypothalamic peptide hormone oxytocin is highly expressed in the bovine corpus luteum: biosynthesis, structure and sequence analysis. EMBO J 3:2351–2354

Khan-Dawood FS, Dawood MY (1983) Human ovaries contain immunoreactive oxytocin. J Clin Endocrinol Metab 57:1129–1132

Khan-Dawood FS, Marut EL, Dawood MY (1984) Oxytocin in the corpus luteum of the Cynomolgus monkey (*Macaca fascicularis*). Endocrinology 115:570–574

Kruip TAM, Vullings HGB, Schams D, Jonis J, Klarenbeek A (1985) Immunocytochemical demonstration of oxytocin in bovine ovarian tissues. Acta Endocrinol (Copenh) 109:537–542

Labhsetwar AP, Collins WE, Tyler WJ, Casida LE (1964) Effect of progesterone and oxytocin on the pituitary – ovarian relationship in heifers. J Reprod Fertil 8:77–83

Lederis K, Jayasena K (1970) Storage of neurohypophysial hormones and the mechanism for their release. International Encyclopedia of pharmacology and therapeutics. Section 41, vol 1. Pergamon, Oxford, pp 111–143
Legros JJ, Franchimont P, Burger H (1975) Variations of neurohypophysial function in normally cycling women. J Clin Endocrinol Metab 91:54–59
Mackenzie K (1911) An experimental investigation of the mechanism of milk secretion with special reference to the action of animal extracts. Q J Exp Physiol 4:305–330
Makino T, Nakazawa K, Ishii K, Haginiwa I, Nakayama A, Iizuka R (1983) Detection of immunoreactive human placental oxytocin and its contractile effect on the uterine muscle. Endocrinol Jpn 30:389–396
Mares SE, Casida LE (1963) Effect of exogenous oxytocin on the progesterone content of the bovine corpus luteum. Endocrinology 72:78–82
McCracken JA (1980) Hormone receptor control of prostaglandin $F_2\alpha$ secretion by the ovine uterus. Adv Prostaglandin Thromboxane Leukotriene Res 8:1329–1344
McCracken JA, Gammal LM, Glew ME, Underwood LF (1978) Hormone receptor control of prostaglandin $F_{2\alpha}$ secretion from the ovine uterus. Endocrinology 102:A440
McCracken JA, Schramm W, Barcikowski B, Wilson L (1981) The identification of prostaglandin $F_{2\alpha}$ as a uterine luteolytic hormone and the hormonal control of its synthesis. Acta Vet Scand [Suppl] 77:71–88
Milne JA (1963) Effects of oxytocin on the oestrous cycle of the ewe. Aust Vet J 39:51–52
Milvae RA, Hansel W (1980) Concurrent uterine venous and ovarian arterial prostaglandin F concentrations in heifers treated with oxytocin. J Reprod Fertil 60:7–15
Mitchell MD, Haynes PJ, Anderson ABM, Turnbull AC (1980) Oxytocin in human ovulation. Lancet II:704
Mitchell MD, Kraemer DL, Brennecke SP, Webb R (1982) Pulsatile release of oxytocin during the estrous cycle, pregnancy and parturition in sheep. Biol Reprod 27:1169–1173
Mukhopadhyay AK, Kumar A, Tams R, Bohnet HG, Leidenberger FA (1984) Oxytocin and vasopressin have no effect on progesterone production and cyclic AMP accumulation by rat luteal cells in vitro. J Reprod Fertil 72:137–141
Neely DP, Stabenfeldt GH, Sauter CL (1979) The effect of exogenous oxytocin on luteal function in mares. J Reprod Fertil 55:303–308
Newcomb R, Booth WD, Rowson LEA (1977) The effect of oxytocin treatment on the levels of prostaglandin F in the blood of heifers. J Reprod Fertil 49:17–24
Nicholson HD, Swann RW, Burford GD, Wathes DC, Porter DG, Pickering BT (1984) Identification of oxytocin and vasopressin in the testis and in adrenal tissue. Regul Pept 8:141–146
Nicholson HD, Peeling WB, Pickering BT (1985) Oxytocin in the prostate and semen? J Endocr 104 Suppl: 127
Ott I, Scott JC (1910) The galactagogue action of the thymus and corpus luteum. Proc Soc Exp Biol 8:49
Ott I, Scott JC (1911) The action of animal extracts upon the secretion of the mammary gland. Ther Gaz 27:689–691
Parry DM, Wilcox DL, Thorburn GD (1980) Ultrastructural and cytochemical study of the bovine corpus luteum. J Reprod Fertil 60:349–357
Pickering BT, Jones CW (1978) The neurophysins. In: Li CH (ed) Hormonal proteins and peptides, vol 5. Academic, New York, pp 103–158
Pickering BT, Birkett SD, Charlton HM, Guldenaar SEF, Nicholson HD, O'Shaughnessy PJ, Swann RW, Wathes DC, Worley RTS (1986) Neurohypophysial peptides in the gonads. Proc 13th International Conference on Biochemical Endocrinology (in press)
Quirk SJ, Wilcox DL, Parry DM, Thornburn GD (1979) Subcellular location of progesterone in the bovine corpus luteum: a biochemical, morphological and cytochemical investigation. Biol Reprod 20:1133–1145
Rice GE, Thorburn GD (1985) Subcellular localization of oxytocin in the ovine corpus luteum. Can J Physiol Pharmacol 63:309–314
Richardson MC, Masson GM (1985) Lack of direct inhibitory action of oxytocin on progesterone production by dispersed cells from human corpus luteum. J Endocrinol 104:149–151

Roberts JS, McCracken JA (1976) Does prostaglandin $F_{2}\alpha$ released from the uterus by oxytocin mediate the oxytocic action of oxytocin? Biol Reprod 15:457–453

Roberts JS, McCracken JA, Gavagan JE, Soloff MS (1976) Oxytocin stimulated release of prostaglandin $F_{2\alpha}$ from ovine endometrium in vitro: correlation with estrous cycle and oxytocin-receptor binding. Endocrinology 99:1107–1114

Robinson AG, Ferin M, Zimmerman EA (1976) Plasma neurophysin levels in monkeys: emphasis on the hypothalamic response to estrogen and ovarian events. Endocrinology 98:468–475

Rowson LE, Tervit R, Brand A (1972) The use of prostaglandin for synchronization of oestrus in cattle. J Reprod Fertil 29:145–154

Schäfer EA, McKenzie K (1911) The action of animal extracts on milk secretion. Proc R Soc Lond B 84:16–22

Schaeffer JM, Liu J, Hsueh AJW, Yen SCC (1984) Presence of oxytocin and arginine vasopressin in human ovary, oviduct and follicular fluid. J Clin Endocrinol Metab 59:970–973

Schallenberger E, Schams D, Bullerman B, Walters DL (1984) Pulsatile secretion of gonadotrophins, ovarian steroids and ovarian oxytocin during prostaglandin-induced regression of the corpus luteum in the cow. J Reprod Fertil 71:493–501

Schams D (1983) Oxytocin determination by radioimmunoassay III. Improvement to subpicrogram sensitivity and application to blood levels in cyclic cattle. Acta Endocrinol (Copenh) 103:180–183

Schams D, Lahlou-Kassi A (1984) Circulating concentrations of oxytocin during pregnancy in ewes. Acta Endocrinol (Copenh) 106:277–281

Schams D, Lahlou-Kassi A, Glatzel P (1982) Oxytocin concentrations in peripheral blood during the oestrous cycle and after ovariectomy in two breeds of sheep with high and low fecundity. J Endocrinol 92:9–13

Schams D, Prokopp S, Barth D (1983) The effect of active and passive immunization against oxytocin on ovarian cyclicity in ewes. Acta Endocrinol (Copenh) 103:337–344

Schams D, Kruip TAM, Koll R (1985) Oxytocin determination in steroid producing tissues and in vitro production in ovarian follicles. Acta Endocrinol (Copenh) 109:530–536

Sheldrick EL, Flint APF (1981) Circulating concentrations of oxytocin during the estrous cycle and early pregnancy in sheep. Prostaglandins 22:631–636

Sheldrick EL, Flint APF (1983a) Ovarian oxytocin and luteal function in the early pregnant sheep. Anim Reprod Sci 10:101–113

Sheldrick EL, Flint APF (1983b) Luteal concentrations of oxytocin decline during early pregnancy in the ewe. J Reprod Fertil 68:477–480

Sheldrick EL, Flint APF (1983c) Regression of the corpora lutea in sheep in response to cloprostenol is not affected by loss of luteal oxytocin after hysterectomy. J Reprod Fertil 68:155–160

Sheldrick EL, Mitchell MD, Flint APF (1980) Delayed luteal regression in ewes immunized against oxytocin. J Reprod Fertil 59:37–42

Swann RW, O'Shaughnessy PJ, Birkett SD, Wathes DC, Porter DG, Pickering BT (1984) Biosynthesis of oxytocin in the corpus luteum. FEBS Lett 174:262–266

Tan GJS, Tweedale R, Biggs JSG (1982a) Effects of oxytocin on the bovine corpus luteum of early pregnancy. J Reprod Fertil 66:75–78

Tan GJS, Tweedale R, Biggs JSG (1982b) Oxytocin may play a role in the control of the human corpus luteum. J Endocrinol 95:65–70

Theodosis DT, Wooding FBP, Sheldrick EL, Flint APF (1985) Ultrastructural localisation of oxytocin and neurophysin in the ovine corpus luteum. Cell Tissue Res (in press)

Walters DL, Schallenberger E (1984) Pulsatile secretion of gonadotrophins, ovarian steroids and ovarian oxytocin during the periovulatory phase of the oestrous cycle in the cow. J Reprod Fertil 71:503–512

Walters DL, Schams D, Schallenberger E (1984) Pulsatile secretion of gonadotrophins, ovarian steroids and ovarian oxytocin during the luteal phase of the oestrous cycle in the cow. J Reprod Fertil 71:479–491

Wathes DC (1984) Possible actions of gonadal oxytocin and vasopressin. J Reprod Fertil 71:315–345

Wathes DC, Swann RW (1982) Is oxytocin an ovarian hormone? Nature 297:225–227

Wathes DC, Swann RW, Pickering BT, Porter DG, Hull MGR, Drife JO (1982) Neurohypophysial hormones in the human ovary. Lancet II:410–412
Wathes DC, Swann RW, Birkett SD, Porter DG, Pickering BT (1983 a) Characterization of oxytocin, vasopressin and neurophysin from the bovine corpus luteum. Endocrinology 113:693–698
Wathes DC, Swann RW, Hull MGR, Drife JO, Porter DG, Pickering BT (1983 b) Gonadal sources of the posterior pituitary hormones. Prog Brain Res 60:513–520
Wathes DC, Swann RW, Pickering BT (1984) Variation in oxytocin, vasopressin and neurophysin concentrations in the bovine ovary during the oestrous cycle and pregnancy. J Reprod Fertil 71:551–557
Watkins WB (1983) Immunohistochemical localization of neurophysin and oxytocin in the sheep corpora lutea. Neuropeptides 4:51–54
Watkins WB, Moore LG, Flint APF, Sheldrick EL (1983) Secretion of neurophysin 1/II and oxytocin by the ovary in sheep. Peptides 5:61–64
Webb R, Mitchell MD, Falconer J, Robinson JS (1981) Temporal relationships between peripheral plasma concentrations of oxytocin, progesterone and 13,14,Dihydro-15-keto prostaglandin $F_{2\alpha}$ during the oestrous cycle and early pregnancy in the ewe. Prostaglandins 22:443–454
Wilks JW (1983) The effect of oxytocin on the corpus luteum of the monkey. Contraception 28:267–272
Worley RTS, Nicholson HD, Pickering BT (1985) Testicular oxytocin: an initiator of seminiferous tubule movement? In: Saez JM, Forest MG, Dazord A, Bertrand J (eds) Recent progress in cellular endocrinology of the testis INSERM, Paris, pp 205–212

Oxytocin and the Milk-Ejection Reflex

I. C. A. F. ROBINSON

Contents

1 Introduction

The most well established biological effects of oxytocin (OT) in female mammals are the stimulation of uterine contractions and the contraction of myoepithelial cells in the mammary gland to promote milk ejection. The physiological role of OT in milk ejection in some species is now well established, and constitutes a prime example of a neuroendocrine reflex (Cross and Harris 1952), the parts of which are now being studied in detail. The role of OT in parturition is much less clear, and varies considerably from species to species. In this brief review I shall summarize the advances that have been made in the role of OT in the milk-ejection reflex over the past few years and hopefully highlight those areas in which further studies are necessary. The role of OT in parturition will not be considered here; the reader is referred to several recent reviews in which this topic is discussed at length (Fuchs F. 1985; Fuchs A. R. 1985; Chard 1985).

Division of Neurophysiology and Neuropharmacology, National Institute for Medical Research, The Ridgeway, Mill Hill NW7 1AA, London, Great Britain, and Malmö Ferring Pharmaceuticals, Box 30562, 20062 Malmö, Sweden

Current Topics in Neuroendocrinology, Vol. 6

2 Methodological Developments

Without doubt, the major advance which has accelerated endocrine studies of OT is the development and application of radioimmunoassays (RIAs) for OT (Chard 1973; Amico et al. 1985). The very low circulating levels of OT in plasma under most circumstances (0.5–2 μU/ml, 1–4 pg/ml 10^{-12} *M*) requires extremely avid antibodies, and the presence of vasopressin (AVP), differing by only two amino acid residues out of nine, requires a highly specific antiserum. Several RIA methods have been developed and used to measure OT in the bloodstream, but as a rule, it has been necessary to combine RIA with an extraction procedure, in order to remove interfering substances, to prevent degradation, or, most importantly, to concentrate the samples to an extent that basal levels become detectable (Khan-Dawood 1985). RIAs have many disadvantages, especially their inability to distinguish biologically inactive but still immunoreactive fragments, but their ability to cope with large numbers of samples outweighs their disadvantages and they have virtually displaced bioassays, except in those cases where the nature of the immunoreactive OT is in question. Despite very sensitive assays being reported, it is still extremely difficult to define basal levels in many small samples (an absolute prerequisite for establishing the secretory profile of a hormone secreted in pulses, with a short half-life), and we can expect further developments in assays to pay dividends. No studies have yet been reported with monoclonal antibodies to OT sufficiently sensitive to measure basal levels, but methods have been developed for identifying neurohypophysial peptides by HPLC after fluorescence derivatization (Gruber et al. 1976), or by direct mass spectroscopy (Flint and Sheldrick 1985). These are far from routine methods at present, and require increases in sensitivity before they can be applied to blood samples. Perhaps an approach such as that adopted in the reverse haemolytic plaque assay (Frawley and Neill 1984) or a cytochemical assay (Chayen et al. 1976) could be constructed with sufficient specificity and sensitivity for OT. It is now well established that OT is synthesized together with a carrier protein, neurophysin (NP), as part of a larger precursor molecule (see chapter by Ivell, this volume) and RIAs for the OT-related NP (NP_{OT}) in several species have been developed, in the hope that the technically simpler NP RIAs could be used to monitor the release of OT indirectly (Dax et al. 1977; Legros et al. 1974; Robinson et al. 1973; North et al. 1980, 1983). Unfortunately, this attractive idea has not borne fruit, and recent studies have shown clearly that NP_{OT} levels are a poor guide to OT secretion, but instead may correlate with an OT-like peptide rather than OT itself (Amico 1985). Until very recently, the posterior pituitary gland was the only known source of OT, and it is widely assumed that the OT found in the bloodstream derives from this neuroendocrine organ. The demonstration that other tissues contain OT-like material, which in the case of the ovary can be released into blood, thereby contributing significantly to the plasma titre of OT (see chapter by Wathes et al., this volume), should be borne in mind when interpreting earlier data on OT levels. However, in the case of lactating animals, the major secretory source of OT would still appear to be the posterior pituitary. One consequence of this is that all stimuli for OT release must activate a neural efferent pathway, and the development of neuronal recordings from cells whose properties allow them to be characterized as OT cells (Lincoln et al. 1973) has opened up the possibility of investigating directly the link between the stimulus (central or via sensory afferents) and the response (OT release and action). The advent of cDNA studies also allows the direct measurement of the biochemical links in the control of OT synthesis. The detection and quantification of the mRNA corresponding to OT will complement our understanding of OT physiology by providing an indirect indication of the capacity for synthesis of the peptide, and how this varies physiologically (Ivell and Richter 1985). One must of course bear in mind that the relationship in time between OT mRNA and the tissue production and rate of export of this peptide remain to be clarified, as does the quantitative relationship between neuronal firing and OT secretion from pituitary terminals.

cDNA technology is also being used to discover the structure and physiology of a number of receptors, and it will be another major step forward when the gene for the OT receptor in mammary tissue is cloned, so that its presence and activity can be studied directly by molecular genetic techniques. The development of methods for studying OT receptor binding (Soloff and Swartz 1973) reminds us that the conversion of an endocrine signal into a biological response is not solely a function of the plasma OT level, but that varying OT receptors levels may play an important part in regulating the function of OT. One can predict that wider availability of high specific activity [^{3}H]OT will direct much greater attention to the physiological role of OT receptors and how they are varied.

3 The Milk-Ejection Reflex

The role of oxytocin in the milk-ejection reflex has in a number of species been well established using bioassays of plasma withdrawn during suckling or milking (Bisset et al. 1970; Cleverley and Folley 1970; Folley and Knaggs 1966), or by cannulating a mammary gland and measuring the increase in milk-ejection pressure produced by contraction of the myoepithelial cells, expelling milk from the mammary alveoli. This is, in effect, an on-line bioassay (Cross and Harris 1952; Lincoln et al. 1973). Afferent stimuli associated with suckling are relayed via the spinal cord to the supraoptic and paraventricular nuclei, and activation of OT neurones (Lincoln and Wakerley 1974, 1975) results in discharge from the posterior pituitary terminals of OT which is then transported in the blood stream to reach the mammary gland, where it triggers milk ejection by contraction of the myoepithelial cells which surround the alveoli (Richardson 1948; Linzell 1955). Although this basic pattern is common to most mammals, the relative importance of OT in the efficient transfer of milk varies from species to species and depends on the architecture of the mammary glands. Those animals with large cisterns allow removal of a sizeable fraction of milk by passive withdrawal by the suckling young, whereas animals whose milk stores are mainly "alveolar" require an active ME response (Lincoln and Paisley 1982). Thus, although the rat may not be typical for most lactating animals, most of the experimental work on the milk-ejection mechanism has been done in this species.

3.1 Rat

In the rat, the litter suckles for most of the day and it has been possible to study the afferent limb of the milk-ejection reflex in some detail, principally because the reflex is not abolished by anaesthesia sufficiently deep to allow the necessary surgical procedures (Wakerley and Lincoln 1973; Lincoln et al. 1973). In a classic series of studies, Lincoln, Wakerley and their colleagues showed that in undisturbed rats, reflex milk ejections (RMEs) occur regularly at 5- to 10-min intervals whilst the mother suckles her litter, and the occurrence of these can be determined simply by observing the "stretch reaction" seen when the pubs vigorously suck to obtain milk (Vorherr et al. 1967). Such movements have also been monitored using a radar system (Lincoln et al. 1980) and shown to coincide with the abrupt onset of an increase in mammary pressure recorded from a cannulated teat. These observations confirm the earlier findings of Lincoln and co-workers (Lincoln et al. 1973; Lincoln and Wakerley 1974, 1975), who related the pup's movements to obtain milk with the synchronized discharge of putative OT neurones recorded in anaesthetized rats during periods of suckling (Fig. 1). The discharge of "OT" neurones always precedes the rise in intramammary pressure and activity of the pubs, showing that the generation of the RME discharge is not simply a reflection of increased sucking activity by the pups immediately prior to the milk ejection response.

The characteristic abrupt rises in intramammary pressure can be mimicked by bolus i.v. injections of 1–2 ng OT (Bisset et al. 1967 a), but not by smaller or larger doses given more slowly or as an infusion (Lincoln et al. 1973). Injection or infusions of much larger amounts of OT produce more extended contractions of the myoepithelial cells, resulting in a series of rhythmic increases in intramammary pressure (Lincoln and Paisley 1982). Such mammary responses are not typical of spontaneous RMEs in the rat, though protracted responses similar to these can be observed under conditions where protracted or large release of OT is induced by other stimuli. For example, lactating rats made tolerant to morphine and then challenged with naloxone (Lincoln and Russell 1985) (Fig. 2) and rats given intracerebroventricular injections of hypertonic saline or carbachol (Clarc and Ro-

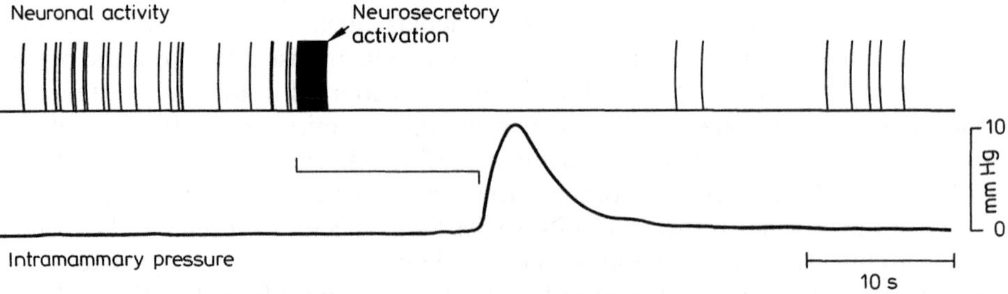

Fig. 1. A recording of the pattern of activity exhibited by a single antidromically identified oxytocinergic neurone during suckling in an anaesthetized lactating rat. Each deflection on the upper trace corresponds to a single action potential. A dramatic increase in spike activity was observed about 12 s before milk ejection, as indicated by the rise in intramammary pressure (*lower trace*). (From Lincoln and Paisley 1982)

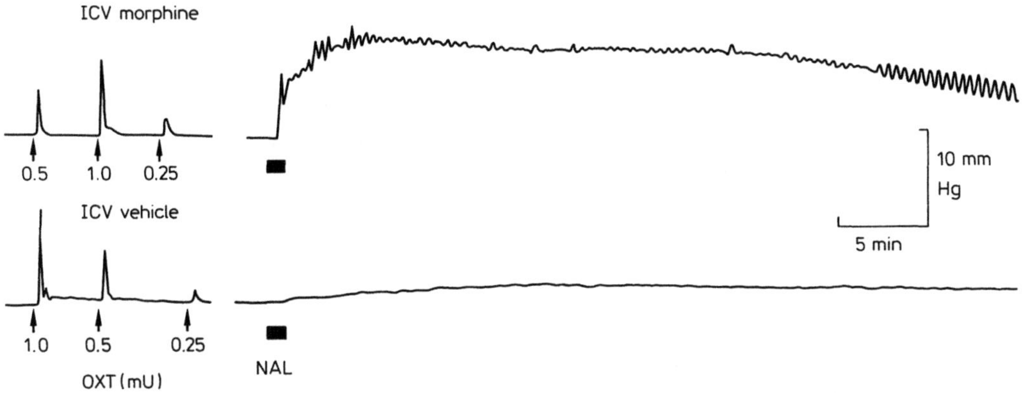

Fig. 2. Intramammary pressure recordings from the cannulated galactophores of lactating rats under urethane anesthesia and previously exposed to 5 days of either morphine (*upper record*) or vehicle infusion (*lower record*) into the lateral cerebral centricles (*ICV*). Opiate withdrawal induced by naloxone (*NAL*) precipitated an immediate, massive release of oxytocin (*upper record*) accompanied by a large and sustained increase in electrical activity of putative oxytocin neurons (not shown). Naloxone had minimal effects in rats not pretreated with morphine (*lower record*). Responses to exogenous oxytocin (*OXT*) were similar in control and morphine-treated rats. (From Lincoln and Russell 1985)

binson, unpublished), release large amounts of OT as measured by RIA, and this elicits protracted intramammary pressure responses. The excellent correlation between onset of spontaneous discharges of identified neurones in the PVN and SON and mammary pressure responses virtually established the identification of OT neurones in terms of functional behaviour, though it has proved difficult until recently to correlate RME discharges with OT measurements in the plasma by independent bioassay or RIA. The development of more sensitive RIAs capable of detecting small amounts of OT in serial samples of small volumes of rat blood has provided confirmatory evidence of OT release of the magnitude envisaged by Lincoln and Wakerley (Higuchi et al. 1985), but the on-line bioassay provided by the maternal mammary gland still provides the most convenient and reliable minute-to-minute monitoring of OT release in the rat. The properties of the milk-ejection reflex in the anaesthetized rat feature some interesting details which are the subject of current studies. The details of the afferent pathway are still being worked out and it is possible to activate small amounts of OT release by electrical stimulation, of the mammary nerves themselves (Mena et al. 1978), though the specificity of this activation is questionable since some ADH release also occurs, The milk-ejection pathway can be traced through the spinal cord and midbrain, but the details of the central pathways for the integration of the afferent signal remain obscure. The milk-ejection reflex is influenced by inputs from several CNS structures, including the septum, hippocampus and subiculum (Tindal and Blake 1984; Lebrun et al. 1983).

One of the most intriguing aspects of the milk-ejection reflex in the rat derives from the neuronal mechanisms necessary to secrete a bolus of OT into the circulation. Close-arterial injection of OT into anaesthetized rats had been developed as one of the most sensitive bioassays for OT still available (Bisset et al. 1967a), with pressure responses recorded to rapid injections of as little at 5–10 pg OT in 100 µl occurring in 3–5 s, indicating that the mammary gland can respond rapidly to exposure to OT. To achieve similar responses, somewhat larger amounts of OT must be injected i.v. and the responses occur at a somewhat longer latency (10 s) after entry to the circulation. If one considers that the discharge of OT neurones occurs around 12 s before the rise in pressure, it is clear that the release of most of the OT must occur during the 1–2 s burst of high-frequency firing, in which the firing rate of the OT neurones can increase to 60–80 Hz from a background, continuous rate of 1–5 Hz. In vitro studies have shown that stimuli delivered at higher frequencies over short periods are much more efficient at releasing OT than the same number of stimulating pulses delivered at a lower frequency over a longer period (Bicknell et al. 1982), suggesting that the episodic electrophysiological patterning of the OT neuronal firing is necessary to achieve sufficiently high plasma concentrations of OT to produce mammary pressure responses. The other requirement for secretion of enough OT in a 2-s burst to produce 1–2 ng/ml in the circulation is rough synchronization of the firing of most, if not all of the OT neurones. Thus, the requirement in the rat for regular large bursts of OT release during suckling has been met by a characteristic electrophysiological patterning of the efferent limb of the reflex in response to a sustained afferent input. There is a tendency to neglect the fact that far more depolarizing spikes reach the OT terminals during the "quiescent" period between milk-ejection responses, and to assume that this OT release must be negligible since no mammary response is observed. However, the on-line mammary bioassay is relatively insensitive and would not detect fluctuations of 1–10 pg/ml in basal OT plasma levels, amounts which in the AVP system span the physiological range of antidiuresis. Studies of the "basal" secretion of OT should be carried out with sensitive RIAs to explore whether the majority of the interburst firing is indeed irrelevant, or may have a more subtle function in releasing small amounts of OT to act at the mammary gland on a more continuous basis, perhaps by regulating the receptor level.

There is nothing particularly unusual in the idea that frequency facilitation provides effective release, but the mechanism responsible for the intercellular synchronization (and even internuclear synchronization) is the subject of both biochemical and morphological studies. In the supraoptic nucleus of the lactacting rat, an ultrastructural reorganization has been observed, allowing a much greater degree of direct contact between magnocellular OT cell bodies and dendrites, with a reduction of glial interposition (Theodosis et al. 1981; Theodosis and Poulain 1984). Richard and his colleagues have sought to implicate a local release of OT within the hypothalamic nuclei as a factor involved in the stimulation and/or synchronization of the excitation of OT neurones. For example, Freund-Mercier and Richard (1981, 1984) showed that i.v. injections of small amounts of OT (10–1000 pg) in anaesthetized lactating rats barely increased the background firing of some but not all putative OT neurones but increased both amplitude and frequency of neurosecretory bursts associated with RME. Importantly, it was not able to induce these bursts in non-suckled rats, indicating a permissive or facilitatory "gating" role rather than a direct excitation. These effects of exogenous OT were blocked by injection of an OT antagonist, and the antagonist itself reduced

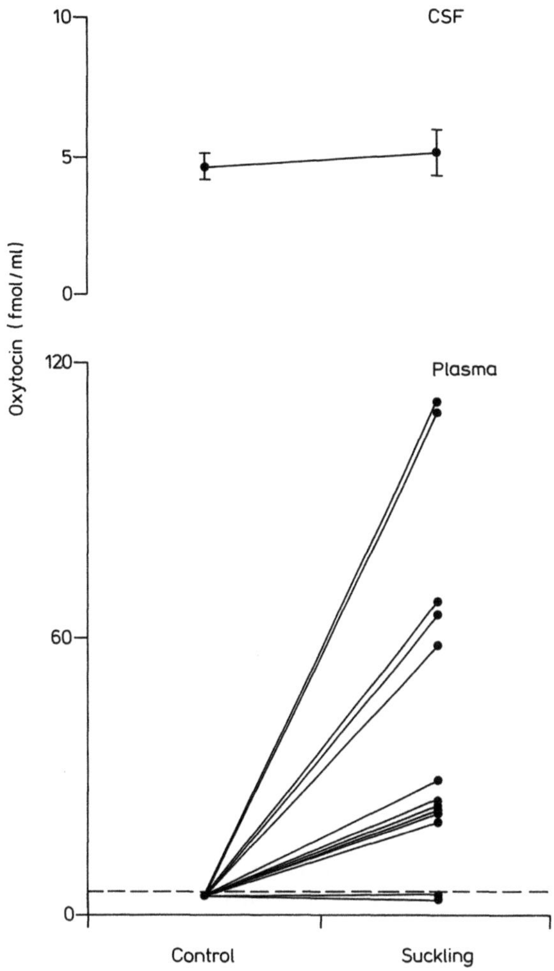

Fig. 3. OT in plasma and CSF samples withdrawn before and at milk ejection in 13 experiments. CSF values are mean ±SEM. The *dotted lines* show the assay threshold. (From Robinson and Jones 1982)

OT release during spontaneous milk ejection, whereas AVP was without effect. The same group have shown that OT can stimulate its own release from pieces of hypothalamus including the magnocellular nuclei, incubated in vitro, postulating that local release of OT may regulate the OT release during the milk-ejection reflex. Robinson and Jones (1982) were unable to detect any changes in bulk CSF OT during suckling in the guinea pig (Fig. 3), but this would not rule out a local release of hormone, confined to small areas of the extracellular space in the vicinity of the magnocellular nuclei (Mason et al. 1986). OT release within the hypothalamus has been suggested to play a role in the onset of maternal behaviour in rats (Pedersen and Prange 1979), but the specificity of this effect is not clear (Pedersen et al. 1982) and the reproducibility has been challenged (Bolwerk and Swanson 1984).

From the foregoing work in the rat, it is apparent that OT plays a central role in achieving an effective milk-ejection reflex, and in this species, a RME is a prerequisite for milk transfer. For example, Kumaresan et al. (1971) and Kumaresan

(1974) gave anti-OT antibodies by passive immunization to lactating rats, and showed that blockade of OT in the mothers reduces milk transfer to the rat pups, as assessed by their weight gain during suckling. In recent years, attention has shifted in an attempt to determine to what extent the basic milk-ejection reflex in the rat has a parallel importance in other species. One of the first electrophysiological differences which emerged concerns the state of CNS arousal during milk ejection. Sleep is a prerequisite for milk-ejection bursts in the rat (Voloschin and Tramezzani 1979; Lincoln et al. 1980), whereas later studies (Poulain et al. 1981) showed that RME was associated with wakefulness in the pig, which otherwise shows a rather similar pattern of abrupt intramammary pressure responses during periods of continuous suckling.

Turning to other species, most of the evidence for the role of OT during milk ejection has relied on hormone measurements, rather than electrophysiological studies, though qualitatively similar neuronal discharges of putative OT cells have been recorded and associated with milk ejection in some species, for example the rabbit (Paisley and Summerlee 1984).

3.2 Rabbit

Rabbits show an extreme version of episodic secretion of OT associated with milk ejection. Cross and Harris (1952) studied nursing behaviour and milk ejection in rabbits, which suckle their young once a day, during which brief period the litter must obtain their total daily milk supply. Not surprisingly, this is reflected in the OT secretory profile. Bisset et al. (1970) used a combination of bioassays on samples of blood withdrawn from conscious rabbits during suckling and demonstrated large amounts of OT (31–375 μμ/ml) in samples taken during successful suckling episodes, whereas little or no AVP (6–24 μμ/ml) was detected. These workers found that the occurrence of OT was consistently related to milk-ejection episodes, but that there was no correlation between the amounts present in a blood sample and the milk yield. Using radioimmunoassay, Fuchs et al. (1984) measured serial blood samples for OT release during early, mid and late lactation in conscious rabbits, and found data essentially confirming the earlier work of Bisset et al. (1970). From rather high basal levels (23 pg/ml) plasma OT values rose sixfold within 1 min of suckling, reaching 200–300 pg/ml 3.5 min after onset and remaining high for 10 or 20 min. These workers also found that the initial peak of OT release was higher in later lactation. Again there was no correlation between OT levels and milk transfer.

Paisley and Summerlee (1984) investigated the magnocellular neuronal responses during RME in conscious rabbits, using implanted microwire recording electrodes, and were able to describe putative OT neurones showing RME-associated bursting discharges, 20 s before the behavioural response of the litter indicated milk ejection. One difference in the rabbit was that further discharges of these OT cells were associated with grooming, suggesting a more complex sensory input to these cells.

In an extension of these studies, it was possible to correlate the OT secretory profile measured in a series of closely timed samples, with the recording of pre-

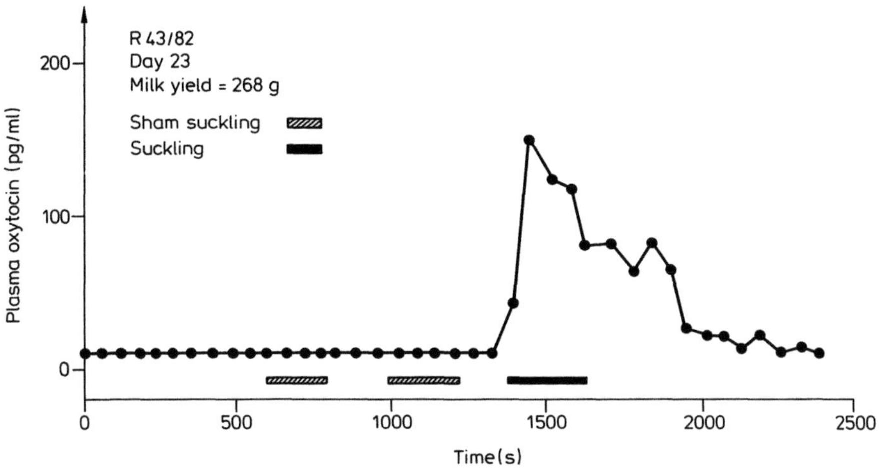

Fig. 4. Blood samples were taken form a conscious rabbit during exposure to the young (*sham suckling*) and then during a period of suckling during which the pups gained 268 g. Measurements of OT by specific RIA showed a large rise in plasma OT only during the real suckling episode. (Data from Summerlee et al. 1986)

sumptive OT neurones (Fig. 4). Although the neuronal recordings are from single cells, and the hormone measurements are affected by distribution and clearance, the neuronal activity and OT secretion closely paralleled each other (Summerlee et al. 1986). Although this is at first sight a gratifying result, it must be remembered that the link between OT neuronal activation and mammary response is susceptible to modification in such a way that one cannot always predict the milk-ejection response from the neuronal firing. It has long been known that the mammary response to OT is inhibited by catecholamines (Bisset et al. 1967 b; Lefcourt and Akers 1984), and we now know that endogenous opioids can inhibit the release of OT from neurohypophysial terminals (Clarke et al. 1979; Clarke and Wright 1984) without necessarily affecting the activity generated at the cell bodies. Both of these types of endogenous modulators may well be released during experimental manoeuvres, such as blood sampling, during suckling in conscious animals.

3.3 Guinea Pig

Guinea pigs also allow only relatively short periods of intense suckling, during which the young obtain appreciable quantities of milk. Robinson et al. (1981) used chronically catheterized animals and a sensitive radioimmunoassay to detect OT release during suckling. As in the rabbit, large amounts of OT were released during a suckling episode, but the levels were variable, depending on the timing of the sample in relation to milk ejection (Fig. 5), and there was no correlation with the amount of milk transferred. In these studies, NP was also measured and was shown to be released in parallel with OT (Fig. 6). In a similar study using pas-

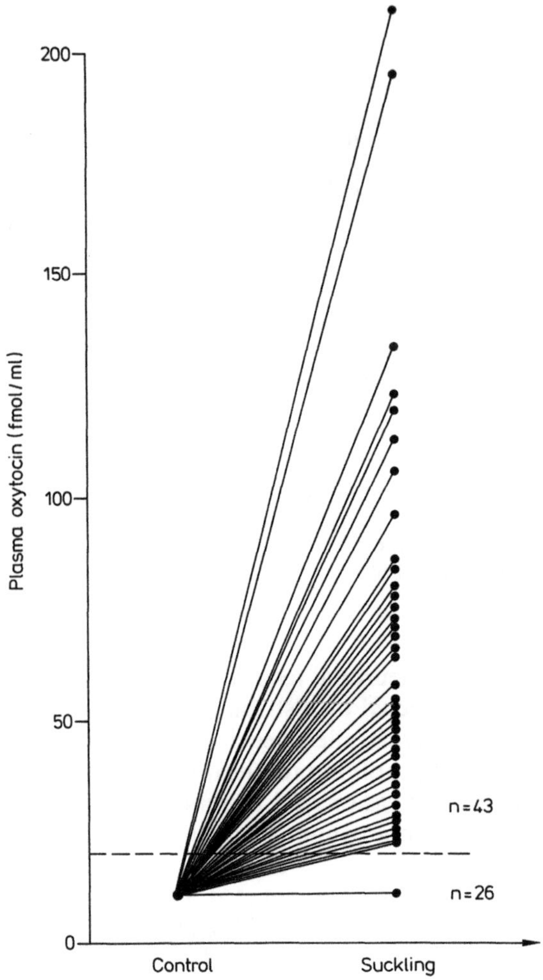

Fig. 5. OT release during suckling in guinea pigs. The *dotted line* shows the detection limit of the least sensitive assay, although most control samples contained less than 5 fmol OT/ml. Suckling samples were withdrawn at or around the time of milk ejection. (From Robinson et al. 1981)

sive immunization against OT it was still possible to observe NP release during suckling, though the immunization reduced, but did not abolish, milk transfer (Fig. 7). Consideration of the time course of OT release and milk ejection together with the rate of binding of OT by anti-OT antisera showed that it is extremely difficult for even a large excess of circulating OT antibodies to mop up enough of the OT secreted, for example in large bursts, before it reaches the mammary circulation to prevent the residual free OT causing a small intramammary pressure rise (Robinson and Parsons 1981). This is a kinetic problem, not one of antibody capacity at equilibrium, a point often ignored in studies attempting to use antisera to block rapidly secreted hormones.

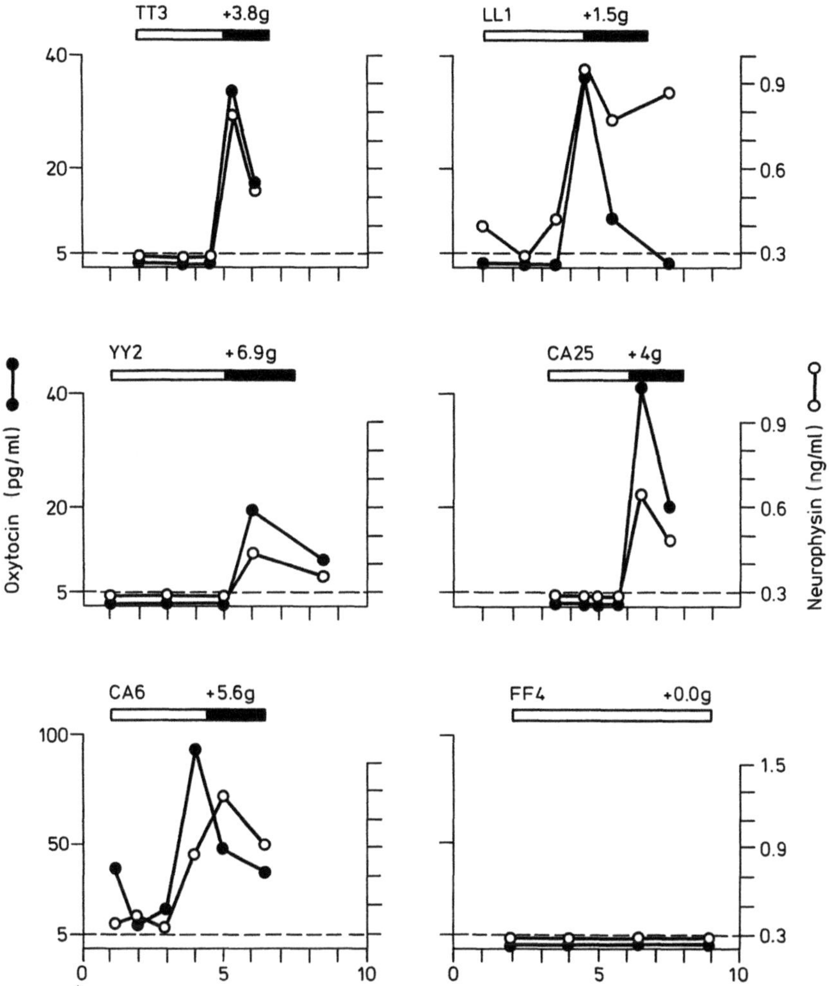

Fig. 6. Typical relationship between plasma levels of OT (●), NP (○) and milk let-down in guinea pigs. The onset of suckling (*open bars*) and of gulping and swallowing (*solid bars*) are shown above each panel, together with the weight gained by the litter in each period. The detection limits of each assay are shown by the *dotted lines*. (From Robinson et al. 1981)

3.4 Other Animal Species

Assay sensitivity is less of a problem in the study of milk ejection in larger animals, since larger blood samples can be obtained and the OT content concentrated in extracts. Both bioassays and RIAs have provided considerably more information about the time course of OT secretion in large species. For example, Cleverley and Folley (1970) were able to detect transient secretory episodes of OT secretion in eight cows during machine milking (9–889 μμ/ml) and found that OT

A

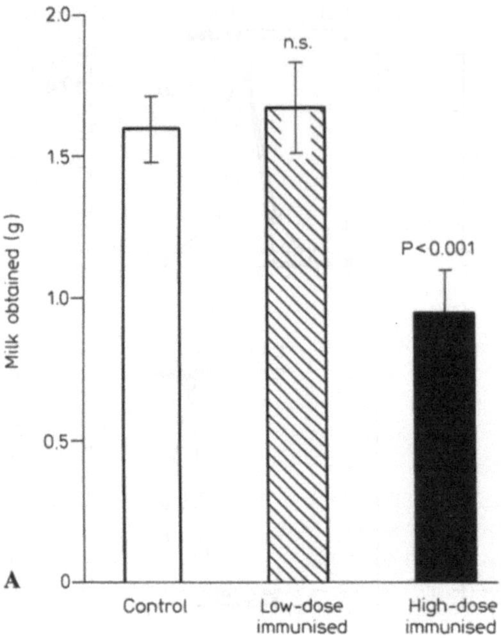

Fig. 7 A. Effect of passive immunization on milk ejection in guinea pigs. The increase (mean ± SEM) in litter weight during suckling periods of 10 min was taken as an estimate of milk transfer. Control group, n = 67 (*open bar*); low-dose group injected with 100 µl OT antiserum, n = 23 (*hatched bar*); high-dose group injected with 400–500 µl OT antiserum, n = 34 (*solid bar*). *NS*, not significant. (From Robinson and Parsons 1981)

B

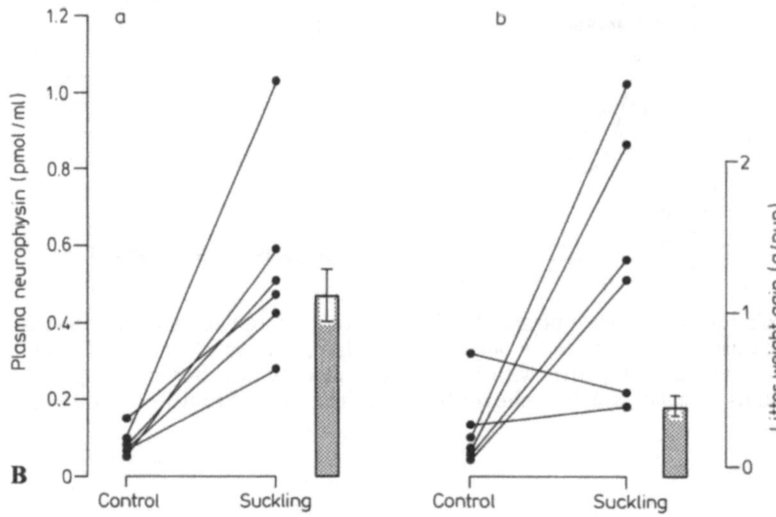

Fig. 7 B. Milk-ejection experiments performed with (*a*) a normal guinea pig and (*b*) one immunized with 400 µl OT antiserum. In both animals, samples taken during suckling experiments showed large rises in plasma levels of NP (●). However, the litter weight gains (mean ± SEM; *open bars*) were much reduced in the six milk-ejection periods in the immunized animal compared with the control animal. (From Robinson and Parsons 1981)

was released effectively by conditioning stimuli associated with the milking routine. The OT-induced rise in milk-ejection pressure was sustained for up to 60 min. The authors suggested that the raised postejection pressure was related to the special morphology of the bovine mammary gland rather than to persistence of OT in the circulation.

Fluctuating OT concentrations can be observed in cows (11–65 μu/ml) during hand or machine milking, with a tendency for the amount of OT release to fall as lactation progresses, and conditioned releases of OT have been observed in response to milking cues, even in animals not in lactation (Gorewit et al. 1983). Despite the large volume of blood (~40 litres) in a dairy cow, as little as 20 mU OT injected as an i.v. bolus can produce a small milk-ejection response. Using a sensitive radioimmunoassay and serial sampling, Schams et al. (1980) found that OT secretion was increased by milking in cows. Although the secretion showed bursts, the length of the burst, as judged by plasma concentrations, was somewhat longer than in the smaller animals.

In the goat, OT release is very variable (5–86 μu/ml) between animals and in different milking periods in the same animal, and McNeilly (1972) was unable to demonstrate a link between the timing of release and any particular aspect of the milking or suckling routine. Although a true conditioned release of OT was not observed, OT was often detected in blood before direct teat stimulation was applied, and OT release was more prominent in early than in late lactation. Since milk yields were normal whether or not OT was detectable, OT release may not be essential for milk ejection in this species.

3.5 Human

Early bioassay data provided evidence that small amounts of OT release could be detected in women during suckling. For example, Coch et al. (1968) obtained internal jugular venous blood whilst monitoring intramammary pressure in lactating women and detected OT ranging from 12 to 25 μu/ml during suckling, which corresponded to the amounts of OT calculated to be produced by infusions of OT at a rate sufficient to elicit milk ejection in lactating women. A number of more recent investigations have used RIAs for OT to study release of this peptide during lactation in women. Piron-Bossuyt et al. (1978) observed spurt release of OT (10–129 μu/ml) in suckling women 4–8 days post partum. These levels were somewhat higher than later investigators found, possibly because of using unextracted blood where oxytocinase may still be effective (Leake and Fisher 1985). The peaks were somewhat random, and there was evidence in some women of OT release 5 min before suckling commenced. Dawood et al. (1981) studied 12 women on the 3rd–5th postpartum day, and found relatively modest rises in plasma OT from 10–50 pg/ml, with wide variations in levels, and noted that some women showed little apparent change in OT levels despite effective suckling. Lucas et al. (1980) used frequent sampling, and found so little evidence of OT release as to conclude that OT release was not necessary for a satisfactory let-down in some women.

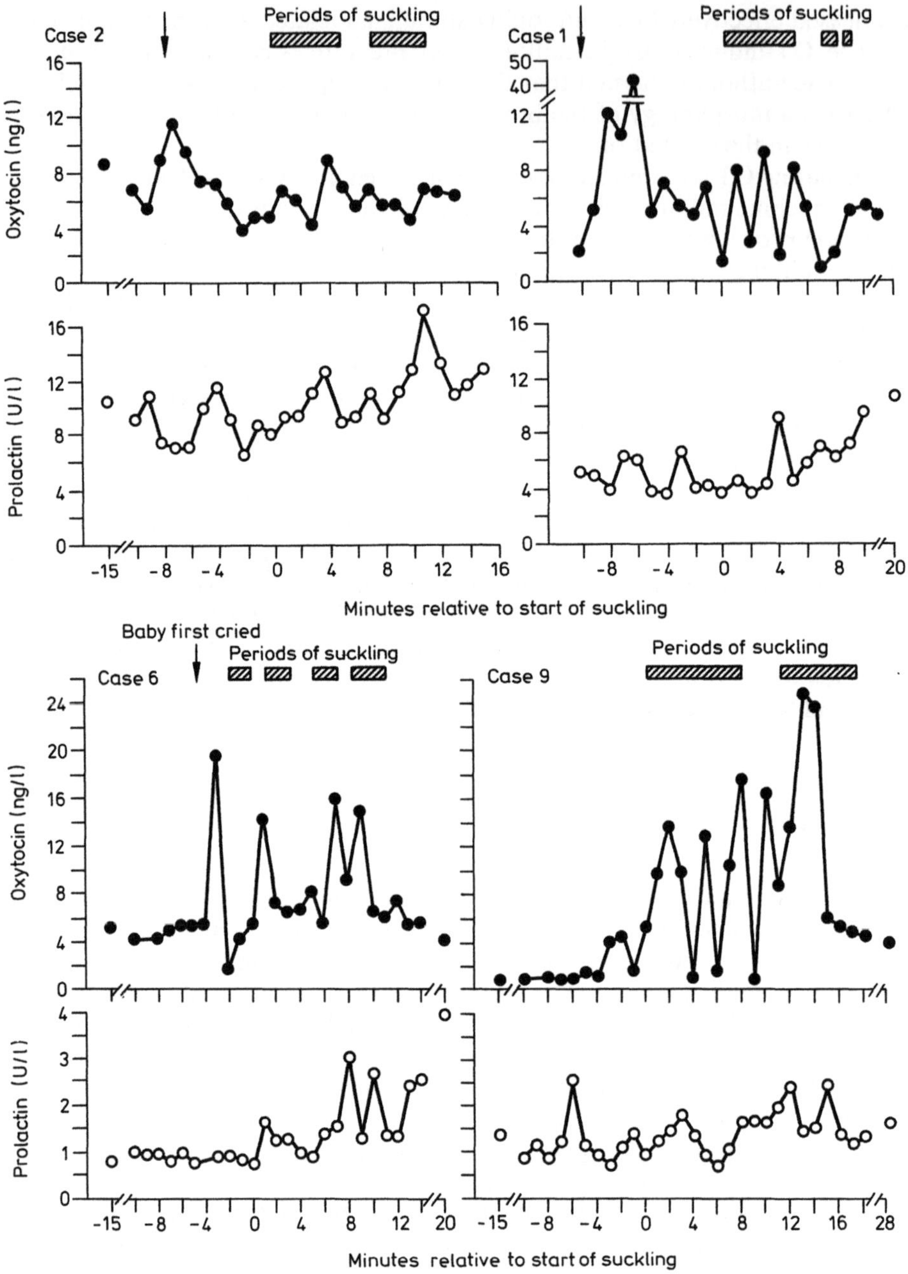

Fig. 8. Changes in plasma OT (●) and prolactin (○) concentrations in women on postpartum day 4 (*cases 1, 2*), day 28 (*case 6*) and day 75 (*case 9*). (From McNeilly et al. 1983)

McNeilly et al. (1983) studied OT release in early and established lactation. Detectable pulses of OT were released in all the women, but the amounts of OT released were relatively small, very highly episodic and seldom exceeded 50 pg/ml from basal values of 1–5 pg/ml. One of the most significant conclusions from this study was that there was often a significant release of OT before the tactile stimulus of suckling in response to such cues as seeing or hearing the baby, which in some cases stimulated more OT release than the subsequent suckling (Fig. 8). McNeilly et al. (1983) concluded that the conditioned releases might be relatively important in women and that stress-induced inhibition of milk ejection might occur immediately prior to suckling, rather than simply during feeding itself. This would go some way to explaining the variability in OT values in earlier studies, if no account had been taken of OT release immediately prior to suckling. Some of the problems experienced by nursing mothers early in lactation may relate to pain associated with breast engorgement, and shifts the focus of attention to a possible inhibition of the effects of OT released in response to the cues of preparation for nursing, rather than directly elicited by suckling at the nipple which produces relatively modest rises of OT during nursing itself (Weitzman et al. 1980).

4 Mammary Gland Receptors for Oxytocin

Oxytocin binding sites (putative receptors) have been demonstrated in the mammary tissue from a number of species. Markle et al. (1978) isolated plasma membrane material from rabbit lactating mammary gland which showed some degree of specific binding to [^{3}H]OT of high affinity (K_d 2 nM). Soloff and his colleagues have studied [^{3}H]OT binding, initially to crude particulate fractions isolated by differential centrifugation (Soloff and Swartz 1973), but also to dissociated cell suspensions enriched in myoepithelial cells, prepared by collagenase digestion, though this procedure produces OT binding sites of somewhat lower affinity (Schroeder et al. 1977; Soloff et al. 1980). The specificity of binding in vitro correlated well with the biological potency of a number of analogues tested, suggesting that the binding sites may well be part of the OT receptor complex, though in common with many peptide receptor studies, the apparent equilibrium dissociation constant was somewhat higher than the concentration of OT necessary for producing a half-maximal biological response in this tissue. One should also remember that these studies are often carried out for much longer periods of time and at different temperatures. Whether or not such studies in broken-cell or digested-cell preparations truly reflect functional OT receptors, they can undoubtedly be used to give interesting information on the functional state of the system. For instance, the properties of the binding sites seem to be different later in lactation (Schroeder et al. 1977), and local variation in OT binding may be observed, suggesting a regional distribution of function within the lactating mammary gland (Soloff et al. 1975). In contrast to the number of studies of OT binding, there is little known about the mechanism of post-receptor activation in mammary myoepithelial cells. Olins and Bremel (1984) have reported that OT stimulates phosphorylation of myosin light chains and that the duration of this effect is related to Ca^{++} influx, which is in line with the well-known dependence of OT-induced mammary contractions on external calcium. These effects were obtained at OT concentrations (10 n*M*) near the range of the K_d for the OT receptor, and were relatively specific for OT. It remains to be seen whether this mechanism exists in myoepithelial mammary tissue from other species.

One major drawback is the lack of OT antagonists with a high specificity towards OT's action on the mammary gland, though there are some anti-uterotonic and anti-vasopressor peptides that show anti-OT activity. Development of analogues suitable for photoaffinity labelling will be a powerful means of progressing towards the characterization of the OT receptor, and the realization that variations in the side chain are tolerated better in antagonists than in agonists may help towards this goal (Sawyer and Manning 1985). Better still, a search for cell lines expressing OT receptors might prove a simpler and ultimately more profitable route to their characterization. From the difference in potency in existing analogues, it is likely that the mammary and uterine receptors will fall into different classes, with other OT receptors in the CNS yet to be classified.

5 Conclusions

Though the involvement of OT in milk ejection is now well established, its importance varies from species to species. The combination of electrophysiological and endocrinological approachs have demonstrated the extraordinary elegance of this neuroendocrine reflex, but only the outlines of the mechanism are known, and many questions require further study. For example, do mammary-type OT receptors exist in other tissues or even in the male? What are the mechanisms for synchronous firing of OT magnocellular neurones, and how do they develop in lactating females? Is OT released centrally to coordinate RME, and through which receptors on which cells does it act? To these questions can be added the problems of the neuronal basis of integration of the continuous suckling input in the rat and of how the reflex is controlled in frequency and amplitude by aminergic and peptidergic influences at both the cell body and terminal level (Richard et al. 1985). Perhaps the answer to some of these questions will shed more light on the physiological role in milk ejection of OT, in some ways still the Cinderella of the established peptide hormones.

References

Amico JA (1985) The radioimmunoassay of oxytocin: new developments. In: Amico JA, Robinson AG (eds) Oxytocin: clinical and laboratory studies. Excerpta Medica, Amsterdam, pp 3–15

Amico JA, Fuchs AR, Haldar J, Leake RD, Robinson ICAF, Robinson AG (1985) Collaborative study of the radioimmunoassay of oxytocin. In: Amico JA, Robinson AG (eds) Oxytocin: clinical and laboratory studies. Excerpta Medica, Amsterdam, pp 44–50

Bicknell RJ, Flint APF, Leng G, Sheldrick EL (1982) Phasic patterns of electrical stimulation enhances oxytocin secretion from the isolated neurohypophysis. Neurosci Lett 30:47–50
Bisset GW, Clark BJ, Haldar J, Harris MC, Lewis GP, Rocha e Silva M Jr (1967 a) The assay of milk-ejecting activity in the lactating rat. Br J Pharmacol Chemother 31:537–549
Bisset GW, Clark BJ, Lewis GP (1967 b) The mechanism of the inhibitory action of adrenaline on the mammary gland. Br J Pharmacol Chemother 31:550–559
Bisset GW, Clark BJ, Haldar J (1970) Blood levels of oxytocin and vasopressin during suckling in the rabbit and the problem of their independent release. J Physiol 206:711–722
Bolwerk ELM, Swanson HH (1984) Does oxytocin play a role in the onset of maternal behaviour in the rat? J Endocrinol 106:353–357
Chard T (1985) Oxytocin: physiology and pathophysiology. In: Baylis PH, Padfield PL (eds) The posterior pituitary. Hormone secretion in health and disease. Dekker, New York, pp 361–390
Chayen J, Daly JR, Loveridge N, Bitensky L (1976) The cytochemical bioassay of hormones. Rec Prog Horm Res 32:33–72
Clarke G, Wright DM (1984) A comparison of analgesia and suppression of oxytocin release by opiates. Br J Pharmacol 83:799–806
Clarke G, Wood P, Merrick L, Lincoln DW (1979) Opiate inhibition of peptide release from the neurohormonal terminals of hypothalamic neurones. Nature 282:746–748
Cleverley JD, Folley SJ (1970) The blood levels of oxytocin during machine milking in cows with some observations on its half-life in the circulation. J Endocrinol 46:347–361
Coch JA, Fielitz C, Brovetto J, Cabot HM, Coda H, Fraga A (1968) Estimation of an oxytocin-like substance in highly purified extracts from the blood of puerperal women during suckling. J Endocrinol 40:137–144
Cross BA, Harris GW (1952) The role of the neurohypophysis in the milk-ejection reflex. J Endocrinol 8:148–160
Dawood MY, Khan-Dawood FS, Waki RS, Fuchs F (1981) Oxytocin release and plasma anterior pituitary and gonadal hormones in women during lactation. J Clin Endocrinol Metab 52:678–683
Dax EM, Cumming IA, Lawson RAS, Johnston CI (1977) The physiological release of specific individual neurophysins into the circulation of pigs. Endocrinology 100:635–641
Flint APF, Sheldrick EL (1985) Ovarian oxytocin. In: Amico JA, Robinson AG (eds) Oxytocin: clinical and laboratory studies. Excerpta Medica, Amsterdam, pp 335–350
Folley SJ, Knaggs GS (1966) Milk-ejection activity (oxytocin) in the external jugular vein blood of the cow, goat and sow, in relation to the stimulus of milking and suckling. J Endocrinol 34:197–214
Frawley LS, Neill JD (1984) A reverse hemolytic plaque assay for microscopic visualization of growth hormone release from individual cells: evidence for somatotrope heterogeneity. Neuroendocrinology 39:484–487
Freund-Mercier MJ, Richard P (1981) Excitatory effects of intraventricular injections of oxytocin on the milk ejection reflex in the rat. Neurosci Lett 23:193–198
Freund-Mercier MJ, Richard P (1984) Electrophysiological evidence for a facilitating control of oxytocin neurones by oxytocin during suckling in the rat. J Physiol 352:447–466
Fuchs AR (1985) Oxytocin in animal parturition. In: Amico JA, Robinson AG (eds) Oxytocin: clinical and laboratory studies. Excerpta Medica, Amsterdam, pp 207–235
Fuchs F (1985) Oxytocin in human parturition. In: Amico JA, Robinson AG (eds) Oxytocin: clinical and laboratory studies. Excerpta Medica, Amsterdam, pp 236–256
Gorewit RC, Wachs EA, Sagi R, Merrill WG (1983) Current concepts on the role of oxytocin in milk ejection. J Dairy Sci 66:2236–2250
Grosvenor CE, Mena F (1979) Alterations in the oxytocin-induced intramammary pressure response after mechanical stimulation of the mammary gland of the anaesthetized lactating rat. Endocrinology 104:443–447

Gruber KA, Stein S, Brink K, Radhakrishnan A, Udenfriend S (1976) Fluorimetric assay of vasopressin and oxytocin. A general approach to the assay of peptides in tissues. Proc Natl Acad Sci USA 73:1314–1318
Higuchi T, Honda K, Fukuoka T, Negoro H, Wakabayaski K (1985) Release of oxytocin during suckling and parturition in the rat. J Endocrinol 105:339–346
Ivell R, Richter D (1985) The oxytocin gene and its expression in the hypothalamus and ovary. In: Amico JA, Robinson AG (eds) Oxytocin: clinical and laboratory studies. Excerpta Medica, Amsterdam, pp 115–123
Khan-Dawood FS (1985) Methods of extraction and concentration of oxytocin for radioimmunoassay. In: Amico JA, Robinson AG (eds) Oxytocin: clinical and laboratory studies. Excerpta Medica, Amsterdam, pp 16–23
Kumaresan P (1974) The effect of oxytocin antibodies on the litter size in rats. Am J Obstet Gynecol 118:68–72
Kumaresan P, Kagan A, Glick S (1971) Oxytocin antibody and lactation and parturition in rats. Nature 230:468–469
Leake RD, Fisher DA (1985) Special considerations in radioimmunoassay measurement of oxytocin in pregnancy plasma. In: Amico JA, Robinson AG (eds) Oxytocin: clinical and laboratory studies. Excerpta Medica, Amsterdam, pp 39–43
Lebrun CJ, Poulain DA, Theodosis DT (1983) The role of the septum in the control of the milk-ejection reflex in the rat: effects of lesions and electrical stimulation. J Physiol 339:17–31
Lefourt AM, Akers RM (1984) Small increases in peripheral noradrenaline inhibit the milk-ejection response by means of a peripheral mechanism. J Endocrinol 100:337–344
Legros JJ, Reynaert R, Peeters G (1974) Specific release of bovine neurophysin I during milking and suckling in the cow. J Endocrinol 60:327–332
Lincoln DW, Paisley AC (1982) Neuroendocrine control of milk ejection. J Reprod Fertil 65:571–586
Lincoln DW, Russell JA (1985) The electrophysiology of magnocellular oxytocin neurones. In: Amico JA, Robinson AG (eds) Oxytocin: clinical and laboratory studies. Excerpta Medica, Amsterdam, pp 53–76
Lincoln DW, Wakerley JB (1974) Electrophysiological evidence for the activation of supraoptic neurosecretory cells during the release of oxytocin. J Physiol 242:533–554
Lincoln DW, Wakerley JB (1975) Factors governing the periodic activation of supra-optic and paraventricular neurosecretory cells during suckling in the rat. J Physiol 250:443–461
Lincoln DW, Hill A, Wakerley JB (1973) The milk-ejection reflex of the rat: an intermittent function not abolished by surgical levels of anaesthesia. J Endocrinol 57:459–476
Lincoln DW, Hentzen K, Hin T, van der Schoot P, Clarke G, Summerlee AJS (1980) Sleep: a prerequisite for reflex milk ejection in the rat. Exp Brain Res 38:151–162
Linzell JL (1955) Some observations on the contractile tissue of the mammary glands. J Physiol 130:251–267
Lucas A, Drewett RB, Mitchell MD (1980) Breast feeding and plasma oxytocin concentrations. Br Med J 281:834–835
Markle HV, Warr JL, Branda LA (1978) Oxytocin binding sites in lactating rabbit mammary gland. Can J Biochem 56:968–976
Mason WT, Hatton GI, Ho YW, Chapman C, Robinson ICAF (1986) Central release of oxytocin, vasopressin and neurophysin by magnocellular neurone depolarisation: evidence in slices of guinea pig and rat hypothalamus. Neuroendocrinology 42:311–322
McNeilly AS, Robinson ICAF, Houston MJ, Howie PW (1983) Release of oxytocin and prolactin in response to suckling. Br Med J 286:257–259
Mena F, Pacheco P, Aguayo D, Clapp C, Grosvenor CE (1978) A rise in intramammary pressure follows electrical stimulation of mammary nerve in anaesthetized rats. Endocrinology 103:1929–1936
North WG, LaRochelle FT Jr, Melton J, Mills RC (1980) Isolation and partial characterization of two human neurophysins: Their use in the development of specific radioimmunoassays. J Clin Endocrinol Metab 51:884–891

North WG, LaRochelle FT Jr, Hardy GR (1983) Radioimmunoassays for individual rat neurophysins. J Endocr 96:373–386

Olins GM, Bremel RD (1984) Oxytocin-stimulated myosin phosphorylation in mammary myoepithelial cells: roles of calcium ions and cyclic nucleotides. Endocrinology 114:1617–1626

Paisley AC, Summerlee AJS (1984) Activity of putative oxytocin neurones during reflex milk-ejection in conscious rabbits. J Physiol 347:465–478

Pedersen CA, Prange AJ Jr (1979) Induction of maternal behaviour in virgin rats after intracerebroventricular administration of oxytocin. Proc Natl Acad Sci 76:6661–6665

Pedersen CA, Ascher JA, Monroe YL, Prange AT Jr (1982) Oxytocin induces maternal behaviour in virgin female rats. Science 216:648–650

Piron-Bossuyt C, Bossuyt A, Van den Driessche R (1978) Détermination de l'oxytocine plasmatique pendant la lactation chez la femme. Annls d'Endocrinologie 39:155–156

Poulain DA, Rodriquez F, Ellendorff F (1981) Sleep is not a prerequisite for the milk ejection reflex in the pig. Exp Brain Res 43:107–110

Richard P, Moos F, Belin V, Freund-Mercier MJ (1985) Aminergic and peptidergic control of oxytocin secretion during suckling. In: Amico JA, Robinson AG (eds) Oxytocin: clinical and laboratory studies. Excerpta Medica, Amsterdam, pp 154–166

Richardson KC (1944) Contractile tissues in the mammary gland, with special reference to myoepithelium in the goat. Proc R Soc Lond [B] 136:30–45

Robinson AG, Archer DF, Tolstoi LF (1973) Neurophysin in women during oxytocin-related events. J Clin Endocrinol Metab 37:645–652

Robinson ICAF (1980) The development and evaluation of a sensitive and specific radioimmunoassay for oxytocin in unextracted plasma. J Immunoassay 1:323–347

Robinson ICAF, Parsons JA (1981) Suckling in the guinea pig: the effects of passive immunization with an antiserum to oxytocin. J Endocrinol 90:237–244

Robinson ICAF, Woolf CN, Parsons JA (1981) Suckling in the guinea pig: the simultaneous release of oxytocin and neurophysin. J Endocrinol 90:227–236

Sawyer WH, Manning M (1985) Design of oxytocin agonists and antagonists. In: Amico JA, Robinson AG (eds) Oxytocin: clinical and laboratory studies. Excerpta Medica, Amsterdam, pp 423–430

Schams D, Schmidt-Polex B, Kruse V (1980) Oxytocin determination by radioimmunoassay in cattle. 1. Method and preliminary physiological data. Acta Endocrinol 92:258–270

Schroeder BT, Chakraborty J, Soloff MV (1977) Binding of [^{3}H] oxytocin to cells isolated from the mammary gland of the lactating rat. J Cell Biol 74:428–440

Soloff MV, Swartz TL (1973) Characterization of a proposed oxytocin receptor in rat mammary gland. J Biol Chem 248:6471–6478

Soloff MV, Rees HD, Sar M, Stumpf WE (1975) Autoradiographic localization of radioactivity from [^{3}H] oxytocin in the rat mammary gland and oviduct. Endocrinology 96:1475–1477

Soloff MS, Chakraborty J, Sadhukhar P, Senitzer D, Wieder M, Fernstrom MA, Sweet P (1980) Purification and characterization of mammary myoepithelial and secretory cells from the lactating rat. Endocrinology 106:887

Summerlee AJS, Lincoln DW (1981) Electrophysiological recordings from oxytocinergic neurones during milk ejection in the unanaesthetized lactating rat. J Endocrinol 90:255–265

Summerlee AJS, Paisley AC, O'Byrne KT, Fairhall KM, Robinson ICAF, Fletcher (1986) Aspects of the neuronal and endocrine components of reflex milk ejection in conscious rabbits. J Endocrinol 108:143–149

Theodosis DT, Poulain DA (1984) Evidence that oxytocin-secreting neurones are involved in the ultrastructural reorganisation of the rat supraoptic nucleus apparent at lactation. Cell Tissue Res 235:217–219

Theodosis DT, Poulain DA, Vincent JD (1981) Possible morphological bases for synchronisation of neuronal firing in the rat supraoptic nucleus during lactation. Neuroscience 6:919–929

Tindal JS, Blake LA (1984) Central inhibiting milk ejection in the rabbit: involvement of hippocampus and subiculum. J Endocrinol 100:125–129
Voloschin LM, Tramezzani JH (1979) Milk-ejection reflex linked to slow-wave sleep in nursing rats. Endocrinology 105:1202–1207
Vorherr H, Kleeman CR, Lehman E (1967) Oxytocin-induced stretch reaction in suckling mice and rats: a semi-quantitative bioassay for oxytocin. Endocrinology 81:711–715
Wakerley JB, Lincoln DW (1973) The milk-ejection reflex of the rat: a 20- to 40-fold acceleration in the firing of paraventricular neurones during oxytocin release. J Endocrinol 57:477–493
Wakerley JB, Dyball REJ, Lincoln DW (1973) Milk ejection in the rat: the result of a selective release of oxytocin. J Endocrinol 57:557–558
Weitzman RE, Leake RD, Rubin RT, Fisher DA (1980) The effect of nursing on neurohypophyseal hormone and prolactin secretion in human subjects. J Clin Endocrinol Metab 51:836–839

Subject Index

MIX
Papier aus verantwortungsvollen Quellen
Paper from responsible sources
FSC® C105338

If you have any concerns about our products,
you can contact us on
ProductSafety@springernature.com

In case Publisher is established outside the EU,
the EU authorized representative is:
Springer Nature Customer Service Center GmbH
Europaplatz 3, 69115 Heidelberg, Germany

Printed by Libri Plureos GmbH
in Hamburg, Germany